THE DEMOCRATIZATION OF COMMUNICATION

Edited by

PHILIP LEE

CARDIFF
UNIVERSITY OF WALES PRESS
1995

© World Association for Christian Communication and University of Wales Press, 1995

British Library Cataloguing in Publication Data

A catalogue record for this book is available from the British Library.

ISBN 0-7083-1323-X (paperback)
       0-7083-1344-2 (cased)

Published on behalf of the
World Association for Christian Communication

Cover design by Olwen Fowler

Typeset at the University of Wales Press
Printed in Great Britain by Dinefwr Press, Llandybïe, Dyfed

THE DEMOCRATIZATION OF COMMUNICATION

To

Michael Traber

journalist, social activist

and colleague

Contents

List of Contributors   ix

1   Introduction: The illusion of democracy   1
    PHILIP LEE

2   The democratic ideal and its enemies   15
    CEES J. HAMELINK

3   That recurrent suspicion: Democratization in a
    global perspective   38
    MAJID TEHRANIAN AND KATHARINE KIA TEHRANIAN

4   Communication ethics as the basis of genuine democracy   75
    CLIFFORD G. CHRISTIANS

5   Democratization of communication as a social
    movement process   92
    ROBERT A. WHITE

6   The journalist: A walking paradox   114
    KAARLE NORDENSTRENG

7   Women and communications technology:
    What are the issues?   130
    COLLEEN ROACH

8   Traditional communication and democratization:
    Practical considerations   141
    PRADIP N. THOMAS

9   The cultural frontier: Repression, violence, and the   153
    liberating alternative
    GEORGE GERBNER

10 Linguistic minorities and the media   173
   NED THOMAS

11 Mass media and religious pluralism   183
   STEWART M. HOOVER

12 Communication: international debate and
   community-based initiatives   197
   CARLOS A. VALLE

Index   217

The Contributors

CLIFFORD G. CHRISTIANS is a Research Professor of Communications at the University of Illinois, Urbana-Champaign, USA, where he directs the doctoral programme in communications. He holds joint appointments as a Professor of Journalism and a Professor of Media Studies. He has a BA in classical philosophy, a BD and Th.M. in theology and culture, an MA in socio-linguistics from Southern California, and a Ph.D. in social communications from Illinois. He has been a visiting scholar in philosophical ethics at Princetown University, in social ethics at the University of Chicago, and a Pew Scholar in ethics at Oxford University. He serves on the editorial boards of a dozen academic journals and is the editor of *Critical Studies in Mass Communication*.

GEORGE GERBNER is Dean Emeritus and Director of the Cultural Indicators research project at the University City Science Center in Philadelphia, USA. From 1964 to 1989 he was Professor and Dean of The Annenberg School for Communication, University of Pennsylvania. He was editor of the *Journal of Communication* and chair of the editorial board of the *International Encyclopedia of Communication* (New York: OUP, 1989). His recent publications include *The Global Media Debate: Its Rise, Fall and Renewal* (Ablex, 1993); *Triumph of the Image: The Media's War in the Persian Gulf. A Global Perspective* (Westview, 1992) and 'Television Violence; the Power and the Peril', in Gail Dinex and Jean M. Humez (eds.), *Gender, Race and Class in the Media; A Critical Text-Reader* (Sage, 1995).

CEES J. HAMELINK is Professor of International Communication at the University of Amsterdam. He was president of the International Association for Mass Communication Research (1990–4). His books include *Finance and Information* (Norwood, NJ: Ablex, 1983), *Cultural Autonomy in Global Communications* (New York: Longman, 1983), *The Technology Gamble* (Norwood, NJ: Ablex, 1988), *Trends in World Communication* (Penang: Southbound, 1994), and *The Politics of World Communication* (London: Sage, 1994).

STEWART M. HOOVER is an Associate Professor in the School of Journalism and Mass Communication and an associate of the Center for Mass Media Research at the University of Colorado at Boulder, USA. He is an internationally known researcher and writer in the area of mass communication and culture. His primary research has focused on the relationship between media and religion in both the North American and European contexts. He is currently Principal Investigator for a three-year project entitled 'Religion in Public Discourse: The Role of the Media', funded by the Lilly Endowment. He is the author of *Mass Media Religion* (Newbury Park, CA: Sage, 1988), co-editor of *Religious Television: Controversies and Conclusions* (Norwood, NJ: Ablex), and co-editor of *Rallies, Rituals and Resistance: Rethinking Media, Religion and Culture.*

PHILIP LEE studied modern languages at the University of Warwick, Coventry, UK, followed by piano and conducting at the Royal Academy of Music, London, from which he graduated with distinction. Since 1976 he has worked for the World Association for Christian Communication at its international office in London, where he is jointly responsible for editing the journal *Media Development* and for implementing WACC's Study and Action Programme. His publications include *Communication For All* (Indore: Satprakashan Sanchar Kendra, 1985 and New York: Maryknoll, revised edition, 1986).

KAARLE NORDENSTRENG is Professor of Journalism and Mass Communication at the University of Tampere, Finland. His Ph.D. is in psychology from the University of Helsinki (1969). He first worked as editor of youth programmes for Finnish radio and later as head of research for the Finnish Broadcasting Company. He has been inter-nationally active at UNESCO as a consultant in communication research and policies (1969–75), in the International Association for

Mass Communication Research (Vice-President 1972–88, and President of the Professional Education Section 1988–) and in the International Organization of Journalists (President 1976–90). He has published seventeen books, monographs and readers and over 300 articles and papers. He was co-editor with Michael Traber of *Few Voices, Many Worlds: Towards a Media Reform Movement* (London: WACC, 1992).

COLLEEN ROACH is Associate Professor at Queens College, City University of New York. She has taught at Fordham University, the New School for Social Research, New York, and Queens University in Canada. She worked for many years at UNESCO in Paris and is widely published both in the US and other countries, particularly in the field of international communications. Her articles have appeared in journals such as *Media, Culture and Society*; *Media Development*; and the *Journal of Communication*. Her books include *Communication and Culture in War and Peace* (Sage, 1993).

KATHARINE TEHRANIAN is currently Senior Fellow at the Center for the Study of World Religions, Harvard University. She is also Assistant Professor of American Studies at the University of Hawaii at Manoa. Her latest publications are *Modernity, Space and Power: The American City in Discourse and Practice* (Cresskill, NY: Hampton Press, 1995) and (with Majid Tehranian) *Restructuring for World Peace: On the Threshold of the 21st Century* (Cresskill, NY: Hampton Press, 1992).

MAJID TEHRANIAN is currently Senior Fellow at the Center for the Study of World Religions and Research Affiliate of the Program on Information Resources Policy, Harvard University. He is also a Professor at the University of Hawaii. His most recent publications include (with Katharine Tehranian) *Restructuring for World Peace: On the Threshold of the 21st Century* (Cresskill, NY: Hampton Press, 1992) and *Technologies of Power: Information Machines and Democratic Prospects* (Norwood, NJ: Ablex, 1990).

NED THOMAS is Director of University of Wales Press and also Director of the Mercator Media Project at University of Wales, Aberystwyth, a European Commission funded research and documentation project on media in minority languages. A former media correspondent for the *Times Educational Supplement* and *Y Cymro*, he was founder editor of the cultural magazine *Planet* and is the author of books on George Orwell, Derek Walcott and Waldo Williams.

PRADIP N. THOMAS completed his doctoral studies at the Centre for Mass Communication Research, University of Leicester, United Kingdom. He was Associate Professor of Communication at the Tamilnadu Theological Seminary, Madurai, India. In 1990 he was appointed Co-ordinator of the Asia Region at the international headquarters of the World Association for Christian Communication, London, where in 1995 he became Director of Studies and Publications.

CARLOS A. VALLE has been General Secretary of the World Association for Christian Comunication, whose international office is in London, since 1986. He was previously Director of the Communications Department of ISEDET, Buenos Aires, Argentina. His publications include *Comunicación es evento* (Río Piedras, Puerto Rico: 1988) and *Comunicación: modelo para armar* (San José, Costa Rica: 1990).

ROBERT A. WHITE SJ studied development sociology and the political economy of Latin America (Ph.D.) at Cornell University, New York, He worked in Honduras from 1975 to 1978 researching popular movements and the use of radio for social development. He was research director of the Centre for the Study of Communication and Culture, London, 1978–88, and is currently director of the Centro Interdisciplinare sulla Comunicazione Sociale at the Gregorian University, Rome. He is also co-editor of the book series 'Communication and Human Values' published by Sage.

1

Introduction: The illusion of democracy

PHILIP LEE

It is a commonplace that 'direct' democracy – the form of government in which the right to make political decisions is exercised directly by the whole body of citizens – does not exist. Today, more than at any time in history, it would be possible by means of communication technologies to link the citizens of a particular community (village, city, nation) in such a way that everyone could participate in decision-making processes at the same time and to the same degree. To do this, everyone would have to have the possibility of discussing and reaching a consensus on a particular course of action; everyone would have to be educated to a minimum basic level; everyone would have to have access to the same information and (for practical purposes) be part of an active multi-channel communication network.

Until recently it was assumed that members of a nation-state shared the same political identity either implicitly, having been born into it, or explicitly having joined it willingly or unwillingly. The collapse and fragmentation of empires, the realities of migration and labour, the reawakening of previously suppressed local identities based on race, language and religion, have combined to question existing notions of national identity. Today it is recognized that there is a plurality of identities which co-exist within particular political and economic frameworks and whose needs are no longer answered by the traditional form of democratic governance. Such a plurality of identities clearly needs a diversity of options which, in some form or other, cohere at the level of political and economic stability.

In place of 'direct' democracy, a seemingly inevitable series of compromises exists. In order to govern, for example, certain people are chosen to represent the vast majority of others and to take decisions on behalf of the collectivity – in most cases this takes no account of

established institutions such as the judiciary, the civil service, the
financial and banking systems, or even the military. For several years at
a time, this élite takes the political and economic decisions which, by
right, belong to the *demos* (people). Thus, democracy becomes a
compromise between the ideal of total participation and the praxis of
delegating responsibility – usually called 'representative' or
'constitutional' democracy. Something is missing:

> If we have learned any lesson from the Western political culture of the past
> ten years, it is that genuine democracy requires more than the election of
> representatives to a legislative assembly in a multi-party system, no matter
> how essential this is. Over and above voting and party politics, democracy
> requires people who can make their wishes known in public and who
> participate in the debate about the type of society and political process
> they aspire to.   (Traber, 1992: 9)

Genuine democracy demands a system of constant interaction with
all the people, accessibility at all levels, a public ethos which allows
conflicting ideas to contend, and which provides for full participation
in reaching consensus on socio-cultural, economic and political goals.
Such interaction takes place at different levels both separately and
simultaneously. According to Majid Tehranian, it takes the form of a
dialogue which, if carried out in freedom and equality of discourse,
leads to 'a successful development of a universal, human civilization
in whose idiom we all need to speak in order to understand the
national and local in their rich variety of human sub-cultures' (see
below, p.72).

Technologically such a form of communication is only beginning to
become possible in the last decade of the twentieth century. The
spurious promise of the 'information superhighway' is that everyone
will have enough relevant information at their fingertips. The reality
is that information is a commodity for which you have to pay, and a
more profound reality is, of course, that access to information is
controlled.

How, then, can people participate and make their wishes known?
The answer lies in making public communication an integral part of
political democracy. Unless people have access to the knowledge they
require, have the education to deal with that knowledge, and are able
to discuss issues in public with their equals in order to influence
actions taken, there can be no genuine participation. The issue is one
of power:

It could be argued that the core of a democratic society is the presence of a public debate about the distribution and execution of power. It is crucial for democratic arrangements that choices made by the power holders are publicly scrutinised and contested. In the public debate, international and cultural products play a significant role. If the interests of the information and culture producers and the powers that be are intertwined, a society's capacity for democratic government is seriously undermined.

(Hamelink, 1994: 92)

Can public scrutiny of decisions that affect the political, social and cultural life of a community ever be more than wishful thinking? For their physical survival people need certain basics such as food, shelter, and health care, which are recognized as human rights. For their social welfare, they need communication. And for their human dignity people need factors that are intrinsic to genuine democracy: reason, responsibility, mutual respect, freedom of expression, and freedom of conscience, all of which are mediated by communication. A prerequisite of democracy, therefore, is the democratization of communication, which in turn requires the empowerment of the individual.

This has profound implications for the structure of society. The history of twentieth-century politics is one of vacillation between autocratic rule, characterized by the concentration of power in a single centre, and non-autocratic rule, characterized by the existence of several centres. Totalitarianism is the most successful contemporary form of autocratic rule and a constitutional democracy is the chief type of non-autocratic government.

The minimal definition of a constitutional democracy is that it should provide for periodic elections with a free choice of candidates; the opportunity to organize competing political parties; voting privileges for adults; decision by majority vote with protection of minority rights; a judiciary that is independent of government; and constitutional safeguards for basic civil liberties. Of course, autocracies often try to legitimize their activities by adopting the constitutional language of, or by establishing institutions similar to, non-autocratic governments. However, in these cases the totalitarian regime simply exercises power through hierarchical procedures that in reality subject everyone to the wishes of the ruling individual or group.

Clearly the ideological attachment of citizens to either an autocratic or a non-autocratic regime is a function of civil education, since it is primarily through systems of education that citizens learn their duties. Education is also the instrument by which governments further the cohesion of their societies and build the kinds of consensus that support their authority. The post-Enlightenment philosophy of 'education for all' went hand in hand with religious dogma, class limitations and the ideological tenets of capitalism, while later socialist experiments with collective education, particularly those in Communist-led countries, imposed severe restrictions on socio-cultural life and consciousness.

Today, in the North, from primary school to postgraduate level, education consists of multi-dimensional studies that take place in a broad cultural environment. In the South, the picture is drastically different. In nine of the world's most densely populated countries, 70 per cent of adult illiterates and more than half of out-of-school children are disempowered through lack of education. The world's first summit on education for all recognized this situation, stating among other things that:

> Education is a pre-eminent means for promoting universal human values, the quality of human resources and respect for cultural diversity . . . The content and methods of education must be developed to enable individuals and societies to play their rightful role in building democratic societies .
>
> (The Delhi Declaration, 1993)

Few would argue with this. However, the historical function of education has been to instil a sense of national identity, ideological and cultural values, to underline the 'difference' between nations. It is only by breaking down such concepts of difference and raising a collective consciousness of a shared humanity that we can begin to appreciate how interdependent human beings are. In this, traditional forms of communication are crucial. In his contribution to this book, Pradip N. Thomas underlines the fact that they play 'a vital role in the process of negotiation that is itself a core element in the self-understanding and growth of traditional communities' (p.150). Yet, through their globalization and consolidation the communication and information industries are ignoring the tension between the traditional and the modern; they are at a remove from the interdependence of the late-twentieth-century world:

> A new critical consciousness is needed, and this can be achieved only by revised attitudes to education. Merely to urge students to insist on one's

own identity, history, tradition, uniqueness may initially get them to name their basic requirements for democracy and for the right to an assured, decently humane existence. But we need to go on and to situate these in a geography of other identities, peoples, cultures, and then to study how, despite their differences, they have always overlapped one another, through unhierarchical influence, crossing, incorporation, recollection, deliberate forgetfulness, and, of course, conflict . . .

The fact is, we are mixed in with one another in ways that most national systems of education have not dreamed of. To match knowledge in the arts and sciences with these integrative realities is the intellectual and cultural challenge of the moment. (Said, 1993: 330–1).

Human interdependence rests on the foundation of dialogue. By means of language the individual is empowered to be in dialogue with other people. Taught by the mother to her child, learned in play and other social activities, reinforced by education and laden with cultural values, language is the key to communication for Plato's 'animal that speaks'. It is the basis of human dignity and the social relationships, knowledge and information on which rest the ability to question, debate and make decisions. That language and literacy are essential to human dignity is affirmed in the Declaration of the Rights of Man and of the Citizen (1789) as well as in the Universal Declaration of Human Rights (1948) both of which endorse the concept of the right to education. More recently the People's Communication Charter (written in preparation for the fiftieth anniversary in 1998 of the Universal Declaration) calls for 'the right to a diversity of languages' and the promotion of 'educational facilities to encourage language learning by all people without discrimination'.

However, the question is not just one of literacy. A person may know both alphabet and figures, and therefore be 'literate', yet at the same time remain 'functionally illiterate'. In this case what the person knows is too diffuse to enable them to participate effectively in the life of society or to find employment in a modern economy. It is arguable that the political and economic élite of some countries have a vested interest in illiteracy, for whom it is a means of domination. Elsewhere it has become a vital factor in any rational strategy to improve socio-economic conditions. As the Declaration of Persepolis (1975) makes clear:

Literacy, like education in general, is not just the driving force of historical change. It is not the only means of liberation, but it is an essential instrument for all social change. (Bataille, 1976: 3)

Education is concerned with the full realization of human potential. Therefore the different levels of education available in a society should respond to people's abilities to discern and define their needs and to act for themselves, thus enabling them to take their own decisions. The second part of Article 26 of the Universal Declaration of Human Rights states that:

> Education shall be directed to the full development of the human personality and to the strengthening of respect for human rights and fundamental freedoms. It shall promote understanding, tolerance and friendship among all nations, racial or religious groups.

This holistic view of the role of education inescapably identifies it with the right to communicate, as pioneered by Jean D'Arcy for whom 'what matters is the establishment or re-establishment of true communication among human beings' (D'Arcy, 1983: xxv). However, the complacency of the developed countries in regard to education and communication ignores the crying needs of the majority of less developed countries. As two sides of the democratic coin, education and communication share the same goal – the empowerment of ordinary people:

> The right to communicate is an ideal that seeks to empower people to participate actively in the search for solutions to problems of development as perceived and defined by them. It means making available to the people the necessary facilities that will enable them to engage in dialogue on an equal footing.   (Ansah, 1992: 56)

For almost three decades people have promoted the right to communicate, and in that time increasing importance has been given to the role of media education and to the media's role in educating people for responsible citizenship. The potential of media education lies in its constantly evolving process of questioning the beliefs and values implicit in the media, of examining critically the ideological and cultural sources of those beliefs and values, and of providing a framework for critical awareness and autonomy which will lead to genuine democratization and participation in society.

The key word here is *potential*. Using media education to this end is an ideal which may be easier to pursue in the North. In contrast, in many countries of the South, media education either does not exist or it is embryonic. In the opinion of certain governments, the role of media education in promoting responsible citizenship is a dubious enterprise since the media are the recognized purveyors of political and cultural systems and of a market ideology that to a large extent maintains the status quo. All the more reason why:

> Successful media education has to involve an empowerment of learners essential to the creation and sustaining of an active democracy and of a public that is not easily manipulable but whose opinion counts on media issues because it is critically informed and capable of making its own independent judgements. (Masterman, 1994: 311)

A second aspect of this argument concerns the professional education of communicators, especially journalists but also those who work in the entertainment industries and in communication education. In the developed countries, the study of mass communications at undergraduate and postgraduate levels, and the professional training of journalists and broadcasters, has long been of the highest quality. Teaching staff, resources and equipment, are adequate while not ideal, and many universities have flourishing research departments. In developing countries the opposite is true. Until recently, students from these countries wishing to continue at postgraduate level frequently came to North America or Europe and studied in the English language. Alternatively they went to the USSR which, until 1989, taught a considerable number of East European, Central Asian, and African students at all levels – but in Russian. A further drawback was that, being a relatively new field of study, the textbooks for mass communications and journalism originated mainly in the USA and Britain:

> In this respect the question of textbooks in communication education can be seen as an issue of cultural emancipation as understood in the debate around the new international information and communication order. At the same time, the textbook problem represents another element of this new order: the need for a better awareness of the cultural and socio-political diversity of the world, whereby communicators should be educated not only to share a national perspective but to pay due attention also to other peoples and ultimately to the international community at large. (Nordenstreng and Traber, 1992: 1)

Education is, of course, a life-long process. Yet perceptions acquired at an early age are difficult to alter. The kind of education (in its broadest sense) on offer, the principles on which it rests, and the social and cultural context in which it is provided, determine the kind of 'cultural environment' promoted by George Gerbner in his chapter. The world's future inhabitants, children and adults, labourers and leaders, seem to be growing up in an atmosphere of insecurity and mistrust over which the media have a profound influence. Gerbner argues authoritatively for 'a cultural environment that is reasonable, free, fair, diverse and non-damaging' (p.170), although such an environment demands notions of concern and respect belied by the present world in which we live.

The decline of fundamental ethical principles

The twentieth century is a century of gross human negligence. Never before have such opportunities existed for understanding and mutual respect between peoples of different social, cultural, religious and ethnic backgrounds. Never before have such technological possibilities existed for communication, co-operation and the advancement of peoples. Instead there have been horrendous wars, tyrannical political systems, massive debt, famine, genocide and human suffering on an unimaginable scale. The authority of traditional sources of moral leadership has evaporated and this spiritual crisis has led to a vacuum and the decline of fundamental ethical principles. Clifford G. Christians, in his chapter, stresses the moral imperative of truth-telling. He argues that it is only 'when truth with significance becomes the defining feature of communication that the global community has the basic resources for peace, solidarity, mutual respect, and equality' (pp.88–9).

The irony is that genuine democracy (or the nearest we can get to it) and genuine democratization of communication rest on those very ethical principles that seem to be disappearing. In the case of multicultural and increasingly secular societies, the role of religion and the place of spiritual values have been called into question and, some would say, replaced by the ethos promoted by the mass media. In the case of the media, although procedural ethics exist, with changing understandings of the meaning and role of public communication and society, there is an urgent need to re-evaluate their significance on the basis of universal norms.

There is no monopoly of ethical values and many people of different cultural, philosophical and religious backgrounds could make distinctive contributions to a theory of fundamental ethical principles of communication. Nevertheless, still conspicuous by its absence is what might be described as an ethical consensus with regard to those values, norms and attitudes without which individuals lose any sense of collective responsibility and fall back on self-interest:

> For the next millennium a way must be found to a society in which men and women possess equal rights and live in solidarity with one another . . . a way must be found to a reconciled multiplicity of cultures, traditions and peoples . . . to a renewed community of men and women . . . to a society in which peacemaking and the peaceful resolution of conflicts is supported . . . to a community of human beings with all creatures in which their rights and integrity are also respected. (Küng, 1990: 67–9)

What are the practical implications of all of this for a genuinely democratic society? Firstly, the state would fulfil its duty to provide public communication which is open to everyone without exception (citizen or refugee, minority or majority, abled or differently abled) all of whom would have equal status. In addition to encouraging commercial market-oriented media, the state would establish and maintain public service media (not just broadcasting), with guaranteed editorial autonomy subject to agreed codes of ethics and to independent critical evaluation. Public service media would set the standards by which the political and socio-cultural life of the country could be measured in terms of human dignity and achievement. In multi-ethnic and multi-cultural societies, one definition of the function of public service media might be their potential for working out social conflicts in constructive and conciliatory ways. Another might be their grounding in ethical responsibility, including maintaining standards of integrity, balance, honesty and impartiality. It would be the duty of a genuinely democratic government to finance public service media precisely because they fulfilled fundamental notions of equality, freedom of expression and social justice.

Secondly, in terms of an enlightened philosophy of education, such a government would provide resources at all levels adequate to the needs of a genuinely democratic society. These would include buildings, equipment and materials and, most importantly of all, professionally trained and equipped teachers. Media education would form an integral part of the curriculum from primary school to university and adult education.

Thirdly, such a government would facilitate access to public service media by non-government organizations, people's movements and social groups. It would create opportunities for ordinary people to create dialogue informally and on their own terms about issues of national and communal interest, actively encouraging a politics of dissent in the interest of a genuine democracy. Cees J. Hamelink makes the point that this is already happening in the form of 'a world political arena in which people in local communities involve themselves directly in the world's problems' (p.34). Evidence of such involvement can be seen in people's movements concerned with, for example, the environment. But the danger, as Robert A. White signals, is when such a social movement joins a hegemonic coalition. 'The longer a movement remains a marginal "prophetic" voice, the more likely it is that it will have time to develop a social ethic of democratic communication' (p.112).

The illusion of democracy lies in the fact that we are constantly told that all these systems are already in place and have only to be accessed. In the privileged countries of the northern hemisphere, political rhetoric about democracy denies the possibility of inequity, inaccessibility and marginalization. 'We hold these truths to be self-evident . . .' rings hollow, echoing with the cries of the outcast and dispossessed. How much more of an illusion for the peoples of Bosnia, Burma, Cambodia, Chechnya, East Timor, Haiti, Iraq, Rwanda, Sudan, and other countries where democracy is a pale shadow of its ideal?

Communication and human dignity

The foundations of human dignity, solidarity and social stability rest on the very democratic values that ought to imbue public education and communication. It is no surprise, therefore, to find issues such as the democratization of communication, the implementation of ethical norms, and the reform of the mass media at the heart of the work of Michael Traber, to whom this book is dedicated. They also underlie the work of the World Association for Christian Communication (WACC) whose philosophy and policies he was instrumental in guiding from 1976 to 1994. Many of the organization's documents reflect his thinking, most notably the *Christian Principles of Communication* (1986) and *Communication and Community: The Manila Declaration* (1989). In addition, he was editor of the international journal *Media Development,* which promotes dialogue between equals, and a founding member of the MacBride Round Table on Communication, whose statements he co-authored from 1989 to 1995.

Michael Traber's philosophy is based on the principles of human dignity, truth-telling and universal solidarity. His 1972 book *Rassimus und weisse Vorherrschaft* (Racism and White Dominance), his many contributions to communication seminars and journals, his editorials in *Media Development,* and his edited book on *The Myth of the Information Revolution* (1986) reflect his concern to identify and promote communication as a human need and a human right.

But communication theory is nothing without communication practice. No amount of talking about ethical principles will help the ordinary person, least of all in the South, unless practical steps for change are taken. In this respect Michael Traber led from the front. As director of Mambo Press in former Rhodesia (1961–70) and managing editor of the newspaper *Moto,* he was expelled by the Smith regime for subversion, and was only welcomed back to the newly independent Zimbabwe in 1980. The intervening years saw him putting theory into practice in many other African countries, notably Zambia, where he was senior lecturer in journalism at the Africa Literature Centre, Kitwe, 1972–6, before taking up the first of several responsibilities in London at the General Secretariat of the WACC. As director of the Periodicals Development Programme he initiated support for 'strategic periodicals' – those that identified with the underprivileged of their countries and which made positive contributions to socio-economic development, justice and peace. In this way newspapers and magazines, such as *The Voice* in South Africa, *Diálogo Popular* in Guatemala, and *Wantok* in Papua New Guinea, were encouraged to play significant roles in highlighting socio-economic and political injustice. As director of the Department of Information and Interpretation, he supervised WACC's various publications and was co-editor with Dr Robert A. White of a series of books on communication and human values, published by Sage and jointly sponsored by WACC and the Centre for the Study of Communication and Culture. He also initiated a series of books on communication by African authors.

But it was as WACC's director of Studies and Publications that Michael Traber introduced the most striking innovations in programme and policy. After WACC's first international congress, held in 1989 on the theme 'Communication for Community', he articulated a Study and Action Programme which aimed at examining and influencing theory and practice in six areas: communication ethics; the right to communicate; communication and religion;

communication, culture and social change; communication education; and women's perspectives. Workshops, seminars, conferences,
new teaching curricula, books and considerable public debate were
the obvious result of this broad-based attempt to integrate the
principles underlying *Many Voices, One World* (UNESCO's 1980
report of the International Commission for the Study of Communication Problems, chaired by Seán MacBride) into the life and work of
WACC, the churches, non-governmental organizations (NGOs) and
other partners in communication. Michael Traber's vision of equality
and justice for all, based on the democratization of communication,
was the impetus for these many activities.

His continual probing of theories and practices, his global search
for new ideas and possibilities, have been an inspiration to many. For
him certain principles are axiomatic:

- *The principle of human dignity.* Human beings have an intrinsic
  and unique value which has to be recognized socially. From this
  stems not just the right to live, but the right to live a life worthy of
  human beings (which is the ultimate rationale of all human
  rights).
- *The principle of freedom.* Deprivation of freedom makes genuine
  communication impossible, and the first sign of repression in all
  societies is usually the curtailment of freedom of speech. The
  silencing of people as a form of punishment, or still worse,
  solitary confinement, are utterly subhuman. But freedom for
  what? Freedom to participate. Freedom to be part of a nation and
  of the human family. Freedom to shape a collective destiny.
- *The principle of truth telling.* Communication is about human
  relationships. All relationships presuppose mutual trust, and the
  basis of such trust is the assumption that we are telling the truth.
  Communication inevitably breaks down when we suspect the
  other of lying.
- *The principle of justice.* Human dignity, freedom and equality are
  values which, when translated into social relationships, produce
  justice, or living-in-justice with all other people. The mass media
  as we know them stand in almost total contradiction to this view.
  They portray the powerful in politics and business, and the stars of
  entertainment and sports. But the poor, the marginalized, the
  refugees, the old, those with disabilities, people of colour,
  children, and, even today, women, are non-people to the media, or

are typecast. Justice in communication is also an international issue. The present global information and communication systems reflect the world's dominant political and economic structures, which maintain and reinforce the dependence of the poorer countries on the richer.

- *The principle of peace.* Violence and war mark the ultimate breakdown of communication, both interpersonal and public. The word is replaced by the gun or the knife. Most wars between nations have started with a series of lies – by governments and the media about the threat of the enemy. If war is the ultimate failure of public communication, peace is its ultimate glory. Peace means people in communication. Peaceful co-existence of peoples with different national, racial and cultural identities, and of different ideological persuasions can, in today's world, only be achieved through communication aimed at conflict resolution. The mass media carry a heavy responsibility in this process.
- *The principle of participation.* Human dignity, freedom, justice and peace: how do we apply these principles to the mass media of today, and make them operational in the decisions leading to the construction of an 'information superhighway' of tomorrow? The answer presupposes a change in direction. Mass and interactive media cannot primarily be considered business enterprises, but are part of the cultural environment in which we live and move. Media, old and new, should contribute to the quality of life of everyone by celebrating all that is genuinely human (Traber, 1994).

Despite studying philosophy and theology, despite ordination, despite gaining a doctorate in mass communications, Michael Traber's instincts remain those of the practising journalist. He followed the 1978–83 debate that led to the establishment of the International Principles of Professional Ethics in Journalism, in which respect for universal values and the diversity of cultures finds a place in Principle VIII. It is an entirely fitting description of the 'true journalist' to whom this book is dedicated:

A true journalist stands for the universal values of humanism, above all peace, democracy, human rights, social progress and national liberation, while respecting the distinctive character, value and dignity of each culture, as well as the right of each people freely to choose and develop its political, social, economic and cultural systems. Thus the journalist participates

2

The democratic ideal and its enemies

CEES J. HAMELINK

The President of the United States and I, believe that an essential prerequisite to sustainable development, for all members of the human family, is the creation of a Global Information Infrastructure. This GII will circle the globe with information superhighways on which all people can travel. The GII will not only be a metaphor for a functioning democracy, it will in fact promote the functioning of democracy by greatly enhancing the participation of citizens in decision-making. I see a new Athenian Age of democracy forged in the fora the GII will create.

(Al Gore, US Vice President, 21 March 1991,
in a speech to the International Telecommunication
Union conference in Buenos Aires)

The speech by which Vice President Gore introduced the Global Information Infrastructure has revitalized a debate that began some sixty years ago. It was in 1932 that Bertold Brecht envisioned radio as a huge participatory communication system linking citizens together (Brecht, 1932). The old debate obviously also revived as a result of the collapse of non-democratic political structures in Eastern Europe and elsewhere, the spread of telematics, the global networking through such instruments as Internet, and processes of globalization in finance and trade.

The debate on communication and democracy has over the past decades addressed two related themes: the democratic organization of public communication and the contribution of public communication to the democratic organization of society. Often these themes have been debated as if they were unrelated. Moreover, the *first* theme has commonly been approached within national frameworks. If one studies the numerous publications on the democratization of communication that have come out in recent years, democracy appears to actively in social transformation towards democratic betterment of society and contributes through dialogue to a climate of confidence in international relations conducive to peace and justice everywhere, to détente, disarmament and national development.

References

Ansah, Paul A. V. (1992). 'The right to communicate: implications for development', in *Media Development* 1/1992.

Bataille, L. (1976) ed. *A Turning Point for Literacy.* Proceedings of the International Symposium on Literacy. Persepolis, Iran, 28 September 1975. Oxford: Pergamon.

D'Arcy, Jean (1983). 'An Ascending Progression' in *The Right to Communicate: A New Human Right*, edited by Desmond Fisher and L. S. Harms. Dublin: Boole Press.

Delhi Declaration, The (1993). Adopted by the leaders of Bangladesh, Brazil, China, Egypt, India, Indonesia, Mexico, Nigeria and Pakistan at the Summit on Education for All, held in New Delhi, India, 13–16 December.

Hamelink, Cees J. (1994). *Trends in World Communication: On Disempowerment and Self-Empowerment.* Penang: Southbound.

Küng, Hans (1990). *Global Responsibility: In Search of a New World Ethic.* Translated by John Bowden, 1991. London: SCM Press.

Masterman, Len (1994). 'Media education and its future', in *Mass Communication Research: On Problems and Policies. The Art of Asking the Right Questions*, edited by Cees J. Hamelink and Olga Linné. Norwood, NJ: Ablex.

Nordenstreng, K. and Traber, M. (1992). *Promotion of Educational Materials for Communication Studies.* Tampere: University of Tampere.

Said, Edward W. (1993). *Culture and Imperialism.* New York: Alfred A. Knopf.

Traber, Michael (1992). 'Communication as a Human Need and Human Right' in *Religion and Society,* Vol. XXXIX No. 1.

Traber, Michael (1994). Proceedings of the Conference on 'Religion, Television and the Information Superhighway'. Annenberg School for Communication, Philadelphia, USA, 22–3 April.

end at the national border. This limited approach does not recognize that without democratization at the world level, the democratic process at the local and national levels is prone to fail (Shiva, 1993: 59). The *second* theme tends to be treated without serious reference to the organization and quality of public communication beyond the concern that communication should be 'free'. The GII project is usually promoted without any critical enquiry into the sort of information that the superhighway will transport. It is arguable, however, that in case the bulk of information consists of commercial and/or pornographic messages, this may not necessarily enhance the participation of citizens in public decision-making. The institutional organization of the GII could be such that an 'Age of Democracy' is forged with the same qualifications as its Athenian example: a highly exclusive arena in which such actors as women or slaves had no place.

The linkage of the organization and the quality of public communication with its contribution to a functioning democracy raises complex questions such as: can the undemocratic sort of capitalism that will finance the GII in fact produce a genuine participatory democratic arrangement? To me this seems highly unlikely and it is at this point that I want to enter the revived debate. Its central themes have to be seen as interlocked. A functioning democracy is in fact a communication process of a special quality. Its basic feature is its interactivity. If the process of public communication in a society is monological, it obstructs the intersubjective discourse without which democracy cannot function. Consequently, we have to reflect on what a democratic arrangement of world communication would look like, since only such an arrangement could contribute to the democratization of world society. I intend to do this by starting from an interpretation of the 'democratic ideal', analysing its implications for world communication and highlighting the impediments to its realization.

One difficulty we encounter in following this approach is that communication science is not very well equipped for the exercise. The field is haunted by a paucity of theoretical reflection, particularly in the area of world communication. The theoretical approaches to world communication we find in the literature are mainly of the empirical-analytical kind. The empirical-analytical approach attempts to provide a coherent descriptive and predictive account of phenomena. It addresses reality in its current manifestation. The limitations of this approach are twofold.

The essential contestability of theory in the social sciences renders the explanatory validity of empirical-analytical theories rather poor. As these theories are characteristically 'underdetermined' (which means that there are always several theoretical perspectives that concur with the empirical observation of reality) the empirical analysis of data provides no arbitration among divergent theories. More important for the present argument, however, is that empirical-analytical theory cannot address such essential issues as: what type of political institutions and practices do we need to meet the challenges of the future? The empirical-analytical approach provides no meaningful guidance for the choices we face. It does not assist the task of public policy-making.

If we want an answer to the question of how world communication should contribute to a democratic arrangement of world politics and how to organize world communication according to the democratic ideal, we need normative theory. The normative approach addresses reality as it should be and provides grounds for normative choices.

In the broader field of the study of world politics, there has always been a strong tradition of normative theorizing which includes thinkers such as Grotius and Kant. Interest in normative international theory became very lively in the 1960s. The ethical problems of nuclear deterrence and the issue of distributive justice in international relations were extensively debated. This was part of a much larger, renewed interest in political philosophy, inspired in part by the example set by John Rawls in his theory of justice (1973). Important normative studies addressed the questions of just and unjust wars (Walzer, 1980), or the international obligation to assist (Singer, 1979). Work by such theorists as Frost (1986), Kratochwil (1989) and Vincent (1986) contributed significantly to new insights in normative international theory. 'The return of normative ethics responded to a sense that there was something badly wrong with the way in which the positivist-empiricist social science of the 1950s formulated the issues and problems of the day' (Brown, 1992: 196).

As Dougherty and Pfaltzgraff (1990: 565) suggest:

> The urgent issues created by the impact of technology upon institutions, the changes in the political environment resulting from ideology and technology, and the implications of popular pressures and demands upon existing political structures will continue to lead students of international relations, and politics more generally, toward a greater interest in normative theory.

In our reflections on world communication and democracy we should expect that a normative approach delivers a description of how a democratic arrangement of world communication should be organized. This approach should also justify this arrangement as superior to alternative arrangements. The latter requirement raises a familiar problem in political and moral philosophy. How can we justify normative choices?

The democratic ideal: Constitutive dimensions

I want to use the concept 'ideal' since it refers to a process. It suggests that we view democracy as a developmental and educational process which is never fully achieved. The concept 'ideal' also refers to a common ground that allows for different modes of implementation in such manifestations as direct, representative, liberal, socialist, bourgeois, or popular democracies. Taking the classical democracy literature (for instance by Schumpeter, Dahl, Pateman or Dworkin) and common political practice in democratic states as points of reference, there is a broad consensus about a definition of the 'democratic ideal' as a political decision-making procedure that enables all those concerned to participate on the basis of equality. This minimalist definition proposes that the fundamental principle of democracy is political equality. Although the comfortable consensus may risk falling apart, I think that this basic procedural definition needs considerable qualification if we want to secure the egalitarian nature of democratic arrangements. As Gould (1988) has convincingly argued, conventional conceptions of democracy propose a system of governance that provides a maximum degree of 'negative' freedom for the governed while it largely ignores the reasoning that full human freedom includes 'positive' freedom. The latter implies that people should be free to exercise their capacity for self-empowerment (Hamelink, 1994a: 142). Following this, the egalitarian nature of democracy implies that all people are equally entitled to the conditions of self-empowerment.

Another problem with the minimalist definition is that political equality conceived in a narrow sense is no guarantee that a democratic procedure enables the widest possible participation of all people in public decision-making. Actually, in most conceptions of democracy only a limited interpretation of people's participation is foreseen. Political equality, however, has meaning only if it goes beyond the right to vote and to be elected. It encompasses civil rights such as freedom of speech. It also extends to institutions through which political equality should be secured. To promote the freedom that is basic to political equality, democratic participation has to extend into areas where ordinary people do not normally participate.

For Pateman, who advocates a normative approach against such 'realists' as Schumpeter and Dahl, participation is 'a process where each individual member of a decision making body has equal power to define the outcome of decisions' (Pateman, 1970: 71). Following this reasoning we first have to extend the standard of political equality to mean the broadest possible participation of all people in processes of public decision-making. Secondly, we have to extend democracy as a decision-making procedure beyond the realm of the political. The democratic process should be moved beyond the political sphere and extend the requirement of participatory institutional arrangements to other social domains. Participatory democracy should therefore apply to policy-making in the sphere of the production, development, and dissemination of information, culture and technology.

This raises the question of how to organize democratic decision-making. There is a strong tendency in most democracies to let a small élite decide on behalf of others. Particularly in large and complex societies, it becomes difficult to avoid forms of delegation of power to politicians, experts, or the forces of the market. There may be nothing wrong with delegating decisions, but those entrusted with deciding for others should provide a full and transparent account to those on whose behalf they act. This implies that a democratic arrangement should have rules, procedures and institutional mechanisms to secure public accountability. The principle of accountability logically implies the possibility of remedial action by those whose rights to participation and equality are violated. Only through effective recourse to remedial measures, can fundamental standards be implemented. If those who take decisions engage in harmful acts, those affected should have access to procedures of complaint, arbitration, adjudication and compensation. The process of establishing the responsibility for decisions taken, and demanding compensation for wrongs inflicted, is essential to the egalitarian nature of the democratic arrangement.

The argument so far assumes a broad consensus about the principle of political equality. On this basis I propose that this principle can only have substance in an egalitarian conception of the democratic ideal. In an egalitarian version of the democratic ideal the constitutive dimensions of a democratic social arrangement are the equality of entitlement to the conditions of self-empowerment, the widest possible participation of all in public decision-making and its extension into all relevant social domains, the establishment of public accountability (for both public and private power holders) and the availability of remedial measures.

The next question to pose is: how can these dimensions be applied to the description of a democratic arrangement of world communication? Within the limits of this chapter the answer can be no more than an approximation in need of much more detail.

World communication and equal entitlement to the conditions of self-empowerment

Among the essential conditions of people's self-empowerment are access to and use of the resources that enable people to express themselves, to communicate those expressions to others, to exchange ideas with others, to be informed about events in the world, to create and control the production of knowledge and to share the world's sources of knowledge. These resources include technical infra-structures, knowledge and skills, financial means, and such natural resources as the electronic spectrum. Their unequal distribution among the world's population obstructs the equal entitlement to the conditions of self-empowerment.

Taking the case of technical infrastructures, we observe that Europe and the USA have almost 70 per cent of the world ownership of radio/TV sets. The African region has 1.3 per cent of the world share of TV ownership. For radio ownership this region's share is 3.7 per cent. Low-income countries with some 55 per cent of the world's population have less than 5 per cent of the world share of telephone lines. In many low-income countries there is less than one telephone line per 100 inhabitants. In contrast: high-income countries have on average fifty telephone lines per 100 inhabitants.

For the distribution of intellectual resources we can take as an indicator the fact that illiteracy hinders one billion adults in their capacity to express themselves. We can also point to the fact that Europe and the USA produce almost 70 per cent of the world's books, that most technological research and development is carried on in a few countries in the North, and that almost all patents for technology are owned by a handful of the world's largest corporations. We can also refer to the increasingly privatized and commercialized production of knowledge. In this process knowledge becomes ever less a common good and more and more the privileged possession of individual owners.

The emerging GATT/WTO (General Agreement on Tariffs and Trade/World Trade Organization) practice on the protection of intellectual property rights reinforces this as it deprives communities of access to their common heritage and transforms it into the exclusive property of private corporations. A case in point is the world's biological systems which are *common human heritage*, but which through technological innovations (biotechnology) are now becoming private property. Unequal access is also found in the area of such natural resources as outer space. Although space resources are defined in international law as *common human heritage*, states and peoples benefit in very different ways from the use of this heritage. There has been an increase in multilateral and bilateral co-operative projects in space, but the consensus is that not all countries have benefited equally.

The developing countries are very concerned that the benefits of space technology should be distributed in fair ways and they have recently re-opened the debate on this in the United Nations Commission on the Peaceful Uses of Outer Space and its Legal Subcommittee. Their preference is the establishment of a formal accord on principles for space technology transfer. The industrialized countries prefer the continuation of the existing practice of international co-operative space projects. Motivated by the discontent of Third World countries over the existing practice in space co-operation, the Legal Subcommittee of Copuos has had on its agenda since 1988 discussion on the possible codification of rules for co-operation in space activities. The basic legal rationale for this discussion is provided by Article 1 of the Outer Space Treaty (1967) which states: 'The exploration and use of outer space, including the moon and other celestial bodies, shall be carried out for the benefit and in the interest of all countries, irrespective of their degree of economic or scientific development, and shall be the province of all mankind.' In the 1992 and 1993 meetings of the Legal Subcommittee there was a basic disagreement on the desirability of a formal legal instrument to enforce space co-operation. The Third World countries claimed the need of a multilateral formal accord that could ensure access to benefits of space technology for all countries. According to the developed countries there was no need for such an instrument and the best way to increase benefits for all countries would be to expand present co-operation programmes on a voluntary basis.

Some observers expect that in time the present disagreement will change towards an agreement on a set of principles for co-operation in space. With the prospect of an agreement there is also the realistic observation that it is unlikely that the need for special treatment for developing countries will be recognized and that the key space players will share benefits equally.

An arrangement of world communication that takes equal entitlement to communication resources seriously, requires far-reaching changes in current political practices in such areas as development assistance, transfer of technology, intellectual property protection, and space co-operation. These practices all reinforce the inegalitarian character of the present world order. Changes would include a drastic increase in overseas development assistance in the field of communication and under conditions more favourable to recipient parties, the adoption of the Unctad Code of Transfer of Technology on the terms proposed by the developing countries, a revision of provisions on the protection of intellectual property in the GATT/WTO multilateral trade accord so as to take the interests of less powerful countries and small producers into account, and the adoption of a multilateral accord on space co-operation and equal benefits.

World communication and the widest possible participation

This principle has two dimensions: it proposes to include all people in decision-making that affects their lives, and to extend such participation beyond the political realm. In the arena of world communication decisions are taken by the bilateral consortium of the most powerful statal and corporate players. In the politics of world communication ordinary people are excluded (Hamelink, 1994b). This is a reflection of world politics in general where, as Muto writes: 'Most of the major decisions which affect the lives of millions of people are made outside their countries, without their knowledge, much less their consent' (Muto, 1993: 156). The current world communication order is highly unrepresentative of the social relations in the world. The present forms of concentration in ownership of communication resources exclude the largest number of people in the world from participating in the control of the world's communication channels.

In the politics of world communication, democratization requires the establishment of a 'transborder participatory democracy' which is no longer based upon the states as key constituents, 'but the people themselves . . . as the chief actors in determining the course of world politics and economics' (Muto, 1993: 156). This will require the establishment of tripartite arrangements for negotiation and decision-making. The idea of involving more than state players in issues of world politics is not new. The International Labour Organization employs the instrument of trilateral negotiations in formulating its policies. Its decision-making conferences are attended by delegations composed of representatives of governments, employers and workers.

The principles of maximum participation and extended equality call for the participation of people in decision-making in former élitist fields such as technology and culture. Especially in the light of the increasing privatization of the production of technical knowledge and cultural expressions, these social spheres should also be subject to democratic control. This requires that affected citizens have a right to participate in decisions about the development and utilization of technology and culture. This is undoubtedly a complex order, since technology and culture are related to special requirements of expertise, skills, and creativity. It will therefore be necessary to explore how to introduce forms of democratic control that do not constrain the essential input of individual expertise and creativity. Individual scientists, engineers, and artists may resist the notion of technological and cultural democracy, but it would be fallacious to believe that they are fully autonomous today. They may stand to gain more than they lose, if their creativity is subject to considerations of common good rather than to corporate profit motives.

World communication and the principle of public accountability

This principle proposes that the key players in world communications should be accountable for making decisions on behalf of others. Decision-making that affects the daily lives of people around the world takes place on matters like the quality of information, the diversity of cultural products, or the security of communications. The decision-makers are increasingly private parties which are neither elected nor held accountable. As a matter of fact the worldwide drive towards deregulation of social domains tends to delegate important areas of social life to private rather than to public control and accountability. Large volumes of social activity are increasingly withdrawn from public accountability, from democratic control, and from the participation of citizens in decision-making.

The global corporations that control ever more facets of people's daily lives, have become less accountable to public authorities everywhere in the world. 'Most corporate leaders, while proudly exercising their constitutionally protected right to influence elections and legislation, deny that they are making public policy merely by doing business. They do not accept responsibility for the social consequences of what they make or how they make it' (Barnet and Cavanagh, 1994: 422). The key issue, therefore, is establishing the public accountability of the most powerful private players. It would seem however that the adoption of strict rules for the public accountability of private players is highly unlikely in the first place and that their enforcement would probably be beyond the power of public authorities. The only effective pressure could come from the main constituencies of these players, the customers in their markets. Ultimately they are dependent upon the people who buy their goods and services and for whom they decide such matters as the quality of food, clothing, entertainment, work, environment or health care. The establishment of accountability, therefore, demands a massive mobilization and politicization of consumer movements around the world.

World communication and remedial action

It is a crucial consideration for the implementation of rights to equal entitlement, participation and accountability that there can be no rights without the option of redress in case of their violation. Rights and remedies are intrinsically related and where no accessible and affordable means of redress are provided, the effective protection of rights is undermined. The old adagium of Roman law states: *'Ubi ius, ibi remedium'*, where there is law there is remedy. One can turn this around and propose that where no remedy is available there is no law. People should be able to seek effective remedy when states or private parties obstruct their democratic rights. This requires limitations on the power of the state as well as a defence against abuses by non-state players. The basis for procedures for individuals and communities to seek redress should be recognition of the formal right to file complaints in cases where public or private actors do not comply with the rules on entitlement, participation and accountability. Obviously such recognition is still no guarantee that all people will effectively enjoy protection of their rights. To improve the chances of compliance with the norms and rules, effective institutional mechanisms are essential.

This means that – as a minimum – an independent committee for monitoring and review needs to be established, as well as a special rapporteur, and an independent tribunal to respond to well-founded complaints through arbitration and adjudication. The independent committee could function much like the existing human rights committees of the United Nations. The members of the committee should be elected and should serve in their personal capacity. The committee members should have experience in the field of human rights and special competence in the different issue-areas of world communication. The committee should appoint a special rapporteur who can conduct independent studies.

The committee should first receive all complaints filed by individuals and communities that may require adjudication by the tribunal. Much like the European Commission of Human Rights, the committee should receive, review and assess complaints about alleged violations of the rules on entitlement, participation, and account-ability. In this way the committee would act as a filter for cases that could be settled out of court and those that must go to the tribunal. The independent tribunal should function like the European Court of Human Rights as a body that is above all parties and that can take binding decisions. Fundamental to this procedure would be a wide recognition of the competence of the independent tribunal to receive complaints from non-state actors, that both individually and collectively have *locus standi* with the court. The opinions of the tribunal should be binding on those who accept its jurisdiction.

Justification

World communication can be arranged in a variety of ways. In a broad sense, the arrangements can be democratic or non-democratic, and both can have a number of distinct manifestations. The current world communication order is largely characterized by disparity and inequity, exclusion and lack of representation, absence of procedures for accountability and redress. By no standard is this a democratic arrangement. If I propose – as an alternative – an egalitarian democratic arrangement of world communication, this suggests a normative preference which is superior to other arrangements. Following the common requirements of normative theory it is necessary to justify this preference.

Herewith we enter into a much debated and contested domain of moral philosophy where different approaches to justification have been developed and applied. According to the Kantian approach the precepts for moral choice can be known through rational thought and are universal; in the tradition of such utilitarians as Bentham or Mill the decisive justification of normative choice is the achievement of social utility, and in social contractarian thought individuals bind themselves voluntarily to a set of principles.

One common feature of these different schools of moral thought is that they consider normative judgements based upon self-interest generally not morally justified. 'From ancient times, philosophers and moralists have expressed the idea that ethical conduct is acceptable from a point of view that is somehow universal' (Singer, 1976: 10). Thinkers as different as Kant, Hume, Bentham, Rawls, Sartre and Habermas 'agree that ethics is in some sense universal' (Singer, 1976: 11). 'They agree that the justification of an ethical principle cannot be in terms of any partial or sectional group' (Singer, 1976: 11).

In order to protect moral choice against partisanship, different methodological approaches have been proposed. For the non-partisanship test Rawls used the original position in which people – behind a 'veil of ignorance' – have no knowledge about their place in society, class position or social status, intelligence or strength etc. Others have used the hypothetical non-partisan observer (utilitarians), the tool of impartial reasoning (Barry), or an 'ideal speech situation' (Habermas). The problem with this methodological requirement is that it implies the possibility of an abstraction from the normal state of partisan perception of reality. This is so remote from human reality that it cannot yield more than a fascinating thought experiment with little meaning for the question of how we organize our political practice.

To Rawls' question, what principles of justice would be chosen in the original position, Habermas added the principle of universalization. 'A norm is justified only if it is "equally good" for each of the persons concerned' (Habermas, 1993: 68). Herewith Habermas proposed the important methodological principle of intersubjectivity. This addresses the potentially undemocratic nature of the justification exercise.

There is in most approaches a tendency towards justification by the specialists, and most – if not all – methods of moral justification, such as Rawls' 'original position', are monological. In Rawls' approach to impartiality 'every individual can undertake to justify basic norms on his own' (Habermas, 1990: 66). However, if the optimal arrangement is to work, there has to be some sort of social consensus about its preferred position. In Habermas's approach to justification, the essence is dialogue among the community of moral subjects. He argues for a communicative, intersubjectivist approach. This is basic to his discourse ethics (Habermas, 1990). 'Whether a norm is justifiable cannot be determined monologically, but only through discursively testing its claim to fairness' (McCarthy in Habermas, 1990: viii). Discourse ethics 'bases the justification of norms on the uncoerced, rational agreement of those subject to them' (McCarthy, 1990: x). 'Only those norms can claim to be valid that meet (or could meet) with the approval of all affected in their capacity as participants in a practical discourse' (Habermas, 1990: 66). Habermas's discursive proposition has strong attractions but its basic flaw is that his method presupposes what should be justified, i.e. a democratic communication arrangement.

The consensus on the need of a universality of ethical principles as basis for justification has run into trouble with the rise of post-modernist thought on moral philosophy. 'Postmodernism tends to claim an abandonment of all metanarratives which could legitimate foundations for truth' (Waugh, 1992: 5). In addition to its cultural-aesthetic manifestations, postmodern thought is mainly a critique of the modernity of the Enlightenment. It is 'a development in thought which represents a thorough-going critique of the assumptions of Enlightenment or the discourses of modernity and their foundation in notions of universal reason' (Waugh, 1992: 3). In a considerable volume of writings postmodernists have denounced the 'grand narratives' of the Enlightenment (Lyotard, 1985). Herewith they reject unitary thought systems, comprehensive moral theories, and the ambition to answer all questions of how to live with one all-encompassing philosophy.

Against this position one can object that it is possible to agree with the denouncing of the absolutisms of past political doctrines and yet recognize that in the liberal pluralist society that the postmodernists champion, 'certain principles are needed if that society is to remain democratic and free' (Wilkin, 1994: 7). With the denial of such Enlightenment principles like justice and equality, one cannot differentiate between principles that are essential to the creation of democratic arrangements and principles that are not. As a result the body of postmodern reading on world politics provides no helpful insights for the future political practices of the world community. postmodern thought leaves the world pretty much as it is.

Postmodernists object to universal moral principles or truths, and favour a pluralism that they want to be tolerated. However, if you exclude normative principles how can you hope to achieve tolerance? Why should we be tolerant? How can tolerance be established without reference to such basic principles as human dignity? The postmodern relativists claim that there can be no universal norms. It remains unclear though how they substantiate this claim. It would seem reasonable to expect that they demonstrate empirically that no universal norms exist.

One of the problems with a relativist position is that there is little hope for justification outside the boundaries of a specific situation. Thus moral relativism may ultimately lead to moral indifference to events beyond the confines of a local scheme of values. Against this, the universalist position accepts that there are values that transcend local boundaries and that those values are applicable to all. Against the postmodernist objections to this position, I would agree with Habermas that the project of the Enlightenment is not quite complete and it is still too early to abandon the world and our common future to moral indifference.

I propose, therefore, that the justification of a normative preference in world politics should be grounded in an argument that demonstrates that the selected normative position is preferable to alternative positions, because it protects more securely than these alternatives a policy principle that is universally valid in world politics. Following this proposition we have to explore whether such a universally valid principle can be identified and whether this is more securely protected by the proposed egalitarian democratic arrangement of world communication than by its alternatives.

The principle that meets the requirement of universal validity is the defence of human rights. As Lukes rightly observes, the principle is accepted virtually everywhere. 'It is also violated virtually everywhere . . . But the virtually universal acceptance, even when hypocritical, is very important' (Lukes, 1993: 20). Human rights currently provide the only universally available set of standards for the dignity and integrity of all human beings. It is in the interest of all people that human rights be respected. There is an international political consensus about human rights. The world political community has recognized the existence of human rights, their universality and indivisibility, and has accepted a machinery for their enforcement. The United Nations World Conference on Human Rights in 1993 reaffirmed that universality, stating in its final declaration: 'The universal nature of human rights is beyond question'. This tells us that international human-rights law represents – however ineffectually – a set of moral claims that is accepted universally and that is worth defending.

I now have to argue that in an egalitarian democratic arrangement of world communication the principle of the defence of human rights is more securely protected than in non-democratic or in non-egalitarian democratic arrangements. At the core of the defence of human rights is a speech situation with specific requirements: the defence of human rights implies that all people can speak up in defence of their own rights or in defence of the rights of others, that reports about rights violations and their perpetrators are not silenced, and that there is public debate about human rights. This demands the absence of all forms of censorship: one of the gravest obstacles for the defence of human rights.

Therefore a communicative arrangement is required in which 'free speech' is more than just a formal, constitutional right. The mere formal recognition does not secure the participation of all in a open and non-monopolized public debate. The free speech situation required for the defence of human rights should provide the social and material conditions for the widest possible participation of all in public debate, secure that all speakers can decide for themselves what to say, that no one is silenced, that speakers accept accountability for what they say and that mechanisms of redress are instituted against all actors who monopolize, constrain or otherwise distort free speech. Non-egalitarian approaches fail to secure this situation since in a variety of ways they permit or even promote forms of speech privilege and censorship. At the same time they fail to provide mechanisms for accountability of those who speak on behalf of others and for remedial measures in cases where speech is deliberately and systematically distorted. Non-egalitarian arrangements cannot secure the absence of censorship by political, economic or other power holders. Authoritarian arrangements adopt censorship as an essential tool of governance. They may formally, even constitutionally, recognize freedom of speech, but this provides no protection against the arbitrary silencing of speech and the monopolistic privilege of selected speech.

A libertarian arrangement would seem a likely protector of free speech since libertarians tend to campaign for anti-censorship causes. The basic flaw of libertarianism, however, is that it is not capable of addressing the tension between liberty and equality. It leaves inequalities outside the political realm untouched, and thereby allows unequal social and economic forces to influence the political process in highly unequal ways thus rendering the political arena a highly uneven playing field. The inequalities ignored by the libertarians reinforce the capacity for oligopolistic control over speech situations. This excludes the majority of people from participation in the public debate and increases the danger of forms of censorship and distortion. As Graber has suggested (1991: 227) 'future civil libertarians must explore the actual relationship between expression and economics.' Certainly, but this relationship can in the end only be meaningfully explored from an egalitarian perspective with its basic notion that economics needs to be organized such as to benefit everybody's liberty.

It cannot be denied that 'civil libertarians from John Dewey to Thomas Emerson have consistently asserted that material resources must be distributed equally if all persons are to have the actual capacity to exercise their expression rights . . .' (Graber, 1991: 219). Thinkers such as Dewey, Addams and Brandeis have 'detailed the ways in which economic disparities inhibited the efficient functioning of the marketplace of ideas' (Graber, 1991: 185). This went far beyond the position of the conservative libertarian school of thought (Burgess, Spencer, and Sumner among others) which in Jeffersonian vein defended a liberty from state intervention. The conservative credo was that government had no business regulating expression or private property. It established a link between speech and property in the sense that 'the system of private property was a necessary, if not a sufficient, condition of a functional system of free speech' (Graber, 1991: 20). As Graber has correctly analysed, the civil libertarians saw the dangers of an unregulated economy but failed to propose adequate rules to address this. Actually, 'they have proposed constitutional doctrines that they admit let wealthy political activists dominate the marketplace of ideas' (1991: 219).

A non-democratic, authoritarian arrangement of world communication would obstruct the defence of human rights. It would reinforce the almost natural inclination of power holders to silence other speakers. It would subsume freedom of speech under limited, sectional interests and would not provide for accountability and redress. A libertarian variant of a democratic world communication order would also constrain free speech. Examples abound of situations in which libertarian societies are unwilling to stop censorship. The much cited case is the far-reaching censorship exercised by US authorities during the Gulf War. 'Censorship is a social instinct' (Smolla, 1992: 4) and an adequate protection against this is more than libertarianism can promise. The only arrangement that produces the speech situation required for the defence of human rights is inspired by egalitarianism.

The enemies of the democratic ideal

> The large private corporation fits oddly into democratic theory and vision. Indeed, it does not fit.
>
> (Charles Lindblom, 1977)

The enemies of the egalitarian democratic ideal are those forces that actively shape the new world order that is currently emerging – largely in response to the collapse of Communism. The new world order poses a serious threat to the project of an egalitarian democracy since it exacerbates existing inequalities and results in a deep erosion of people's liberty to achieve self-empowerment. Since the new world order is not welcome everywhere, it also provokes a fierce opposition in forms of national, ethnic and religious fundamentalism that – ironically – equally threaten the prospect of an egalitarian democratic arrangement of world communication. The new world order is characterized by the following features.

- It is driven by a 'globalization-from-above' (Falk, 1993: 39) which is controlled by the world's largest business corporations, the most powerful industrial states and their political and intellectual élites, often with the generous support of the media moguls of the late twentieth century. The global reach of these forces is not matched by their acceptance of global responsibility. In fact, the most salient characteristic of the new world order is an ever wider ranging control over people's daily lives without even minimal public accountability.
- The proponents of the new order readily claim that the process is democratic and supports global harmony and prosperity. The

promotional language suggests that a free global market represents people's best interests. The promise that global trading in a deregulated global market leads to unprecedented prosperity for all, does not explain why in the real world the development of prosperity is highly uneven; why in fact poverty and inequality are worsening; why of the 5.6 billion people in the world at the end of 1994 over one billion try to survive on less than US$ 370 per year; why one billion adults are illiterate; why for over 500 million children there are no schools; why of the 2.8 billion labour force over 30 per cent is unemployed. A significant characteristic of the new world order is that 'far from producing a solution to the gap between the world's 'haves' and 'have-nots', the changing structures of international business and investment may exacerbate them' (Kennedy, 1993: 47).

- The globalization of the new world order is characterized by fragmentation and social Darwinism; the leading economic theory suggests that poor countries 'only become relevant when they learn the lessons of the marketplace and possess those features which allow them to compete in the borderless world' (Kennedy, 1993: 61). This is corroborated by a social Darwinism that suggests that those who cannot make it in the market-place have basically themselves to blame for their own inadequacies.

- In several parts of the world, globalization-from-above has already led to aggressive forms of rejection of this imposition of modernity; the new world order is based upon the assumption that the poor will emulate the example of the rich and that they are eager to learn how to consume. It does not cater for the very real possibility that the poor may entrench into forms of national and religious fundamentalism and reject a world order that teaches them to want more material possessions or to move in large numbers to the places of prosperity.

- The new world order combines an unprecedented concentration of power with a stunning parochialism. 'The G-7, the group of the seven most powerful countries, dictate global affairs, but they remain narrow, local, and parochial in terms of the interests of all the world's communities. The World Bank is not really a bank that serves the interest of all the world's communities' (Shiva, 1993: 54).

The emergence of the new world order is greatly facilitated by a conservative libertarian belief system that is broadcast widely across the globe by the world's largest communication conglomerates. Among the essential beliefs that legitimize the opposition to the democratic ideal are the following.

A key belief is the gospel of privatization. It declares that the world's resources are basically private property, that public affairs should be regulated by private parties on free markets, and that the state should retreat from most – if not all – domains that affect people's daily lives. This conventional belief that a free market guarantees the optimal delivery of ideas and information 'has recently enjoyed a spirited revival as the unquestioned creed of the new fundamentalists of privatisation' (Murdock, 1994: 3).

An equally important belief is that 'the very essence of the democratic process' is 'the freedom to persuade and suggest' (Chomsky, 1989: 16). This freedom to 'engineer consent' may be concentrated in the hands of a few social actors only, but this should not be a concern in a free society and certainly provides no rationale for interfering in any way with the activities of the small oligarchies of business that control the provision of information and production of culture.

Very helpful to the enemies of the democratic ideal also is the widespread belief that people cannot be trusted to make sound and sensible decisions about their own lives. People can be left free to select what they eat and wear, but their choice of a system of governance is better left to those in control and their allies in engineering consent.

Conclusion

It is inevitable that this essay concludes by questioning the realistic prospect of an egalitarian democratic arrangement of world communication. Given the formidable power of the driving forces of the new world order and the supporting belief system, this prospect would seem rather dim, to put it generously. It seems unlikely that we could mobilize counterforces against a world order which provides uneven access to the world's communication resources and which reinforces a growing gap between information-rich and information-poor nations and individuals.

Even so, one could argue in support of such mobilization that the current arrangement can only continue as long as people believe that a world communication order which is exclusive, inequitable and unresponsive, is in their best interest. Since the democratization of the world communication order is not in the interest of current power holders we have to bank upon the realization of a 'globalization-from-below' against the 'globalization-from-above' (Falk, 1993: 39). The key component of an egalitarian democratic arrangement is a vibrant, active, self-mobilizing world civil society. Today this is no longer a chimera:

> Twenty years ago many despaired that global problems were spinning out of control. But slowly, inexorably, communities have shown that global change is within their power. They have cut the world's problem down to manageable size and exerted influence far in excess of their numbers. They have ended wars, freed political prisoners, cleaned up the global environment, rebuilt villages, and restored hope.   (Shuman, 1994: 91)

Millions of people around the world are involved with forms of local community-based activities that focus on global problems. A new type of world politics emerges through the proliferation of such initiatives. It represents a shift from conventional international relations mainly conducted by the national foreign affairs élites of statesmen, diplomats, and politicians towards a world political arena in which people in local communities involve themselves directly in the world's problems, often bypassing their national officials. As these local communities begin to network and co-operate, a new formidable force in the shaping of world politics develops. Local communities no longer depend upon the national leadership to make the world a safer place to live. In this process, the incorporation of the local community in a globalization-from-above, is countered by local communities that initiate a globalization-from-below.

Local communities have begun to recognize responsibility for problems outside their boundaries and have put world problems on their policy agenda. Local involvement in global affairs provides people with the opportunity to address this responsibility and increases people's contribution to political life. In this process, people in local communities accept that the realization of the democratic ideal implies a fundamental obligation to take the future into their own hands. As local communities around the world are presently engaged in such areas of activity as development, environment, and human rights, it could be argued that the achievement of a democratic world communication order should equally be put on their agenda as a decisive contribution to the quality of life in the third millennium.

The forces behind the new world order and their fundamentalist opponents divide our planet in endless repetitions of 'us' versus 'them' conflicts. Against this the most effective remedy is to achieve a level of distance from our own sectional interests that allows us to see 'everyone's life as of equal worth and everyone's well-being and freedom as equally valuable' (Lukes, 1993: 36). World political reality is not very encouraging for those who adopt this egalitarian perspective. But then, unless one is beyond caring about our common future, there is no other sensible perspective available.

References

Ashley, D. (1990). 'Habermas and the completion of "The Project of Modernity"', in B. S. Turner (ed.), *Theories of Modernity and Postmodernity*. London: Sage. pp. 88–107.

Baker, J. (1987). *Arguing for Equality*. London: Verso.

Barnet, R. J. and J. Cavanagh (1994). *Global Dreams. Imperial Corporations and the New World Order*. New York: Simon & Schuster.

Brecht, B. (1932). *Theory of Radio*. Gesammelte Werke, Band VIII.

Brown, C. (1992). *International Relations Theory. New Normative Approaches*. New York: Harvester.

Chomsky, N. (1989). *Necessary Illusions*. Boston: South End Press.

Curran, J. (1991). 'Mass media and democracy: a reappraisal', in J. Curran and M. Gurevitch, *Mass Media and Society*. London: Edward Arnold, pp. 82–117.

Dahl, R. A. (1956). *A Preface to Democratic Theory*. Chicago: University of Chicago Press.

Dahl, R. A. (1989). *Democracy and its Critics*. New Haven: Yale University Press.

Dahlgren, P. and C. Sparks (eds.) (1991). *Communication and Citizenship*. London: Routledge.

Der Derian, J. and M. J. Shapiro (eds.) (1989). *International/Intertextual Relations: Postmodern readings of world politics*. Lexington, Mass.: Lexington Books.

Dougherty, J. E. and R. L. Pfaltzgraff Jr. (1990). *Contending Theories of International Relations*. New York: HarperCollins.

Dworkin, R. (1985). *A Matter of Principle*. Cambridge, Mass.: Harvard University Press)

Falk, R. A. (1983). *The End of World Order: Essays on Normative International Relations*. New York: Holmes and Meier.

Falk, R. A. (1993). 'The making of global citizenship', in J. Brecher, J. Brown Childs and J. Cutler (eds.), *Global Visions. Beyond the New World Order*. Boston: South End Press. pp.39–50.

Frost, M. (1986). *Towards a Normative Theory of International Relations*. Cambridge: Cambridge University Press.

Golding, P. (1990). 'Political communication and citizenship: the media and democracy in an inegalitarian social order', in M. Ferguson (ed.), *Public Communication. The New Imperatives*. London: Sage, pp. 84–100.

Golding, P. (1994). 'The communications paradox: Inequalty at the national and international levels', in *Media Development* (XLI), 4: 7–9.

Gould, C. C. (1988). *Rethinking Democracy. Freedom and Social Cooperation in Politics, Economy, and Society*. Cambridge: Cambridge University Press.

Graber, M. A. (1991). *Transforming Free Speech*. Berkeley: University of California Press.

Habermas, J. (1993). *Moral Consciousness and Communicative Action*. Cambridge (Mass): The MIT Press.

Hamelink, C. J. (1994a). *Trends in World Communication*. Penang: Southbound Publishers.

Hamelink, C. J. (1994b). *The Politics of World Communication*. London: Sage.

Hare, J. E. and C. B. Joynt (1982). *Ethics and International Affairs*. New York: St Martin's Press.

Hills, J. with S. Papathanassopoulos (1991). *The Democracy Gap. The Politics of Information and Communication Technologies in the United States and Europe*. New York: Greenwood Press.

International Commission for the Study of Communication in Society (1980). *Many Voices, One World*. Paris: Unesco.

Keane, J. (1991). *The Media and Democracy*. Cambridge: Polity Press.

Kennedy, P. (1993). *Preparing for the Twenty-First Century*. New York: Vintage Books.

Kratochwil, F. V. (1989). *Rules, Norms, and Decisions: On the Conditions of Practical and Legal Reasoning in International Relations and Domestic Affairs*. Cambridge: Cambridge University Press.

Lichtenberg, J. (ed.) (1990). *Democracy and the Mass Media*. Cambridge: Cambridge University Press.

Lindblom, Ch. E. (1977). *Politics and Markets: The World's Political-Economic Systems*. New York: Basic Books.

Lukes, S. (1993). 'Five fables about human rights', In S. Shute and S. Hurley (eds.), *On Human Rights*. New York: Basic Books, pp. 19–40.

Lyotard, J-F. (1985). *The Postmodern Condition*. Manchester: Manchester University Press.

Manaev, O. and Pryliuk, Y. (1993). *Media in Transition: From Totalitarianism to Democracy*. Kiev: Abris.

Murdock, G. (1994). 'The new Mogul empires: media concentration and control in the age of divergence', in *Media Development* (XLI), 4: 3–6.

Muto, I. (1993). 'For an alliance of hope', in J., Brecher, J. Brown Childs and J. Cutler (eds.), *Global Visions. Beyond the New World Order*. Boston: South End Press, pp. 147–62.

Pateman, C. (1970). *Participation and Democratic Theory*. Cambridge: Cambridge University Press.

Raboy, M. and P. A. Bruck (1989). *Communication For and Against Democracy*. Montreal: Black Rose Books.

Raboy, M. and B. Dagenais (1992). *Media, Crisis and Democracy*. London: Sage.

Rawls, J. (1973). *A Theory of Justice*. Oxford: Oxford University Press.

Schumpeter, J. A. (1942). *Capitalism, Socialism and Democracy*. London: Allen & Unwin.

Shiva, V. (1993). 'The greening of the global reach', in J., Brecher, J. Brown Childs and J. Cutler (eds.), *Global Visions. Beyond the New World Order*. Boston: South End Press, pp. 53–60.

Shuman, M. (1994). *Towards a Global Village. International Community Development Initiatives*. London: Pluto Press.

Singer, P. (1979). *Practical Ethics*. Cambridge: Cambridge University Press.

Smolla, R. A. (1992). *Free Speech in An Open Society*. New York: Alfred A. Knopf.

Splichal, S. and J. Wasko (eds.) (1993). *Communication and Democracy*. Norwood: Ablex.

Vincent, R. J. (1986). *Human Rights and International Relations*. Cambridge: Cambridge University Press.

Viotti, P. R. and M. V. Kauppi (1993). *International Relations Theory. Realism, Pluralism, Globalism*. New York: Macmillan.

Walzer, M. (1980). *Just and Unjust Wars: A Moral Argument with Historical Illustrations*. Harmondsworth: Penguin.

Waugh, P. (ed.) (1992). *Postmodernism*. London: Edward Arnold.

Wilkin, P. (1994). 'In need of truth and principles', in *Index on Censorship* (23), 6: 6–7.

Williams, F. and J. V. Pavlik (eds.) (1994). *The People's Right to Know. Media, Democracy, and the Information Highway*. Hillsdale NJ: Lawrence Erlbaum Associates.

3

That recurrent suspicion:
Democratization in a global perspective

MAJID TEHRANIAN AND KATHARINE KIA TEHRANIAN

> Democracy is a recurrent suspicion that more than half of the people are right more than half of the time. (E. B. White)
>
> Democracy is government by discussion. (Sir Ernest Barker)
>
> . . . that the government of the people, by the people, for the people shall not perish from the earth. (Abraham Lincoln)

This chapter reconceptualizes the problem of democratization in terms of the bottom-up responses to the top-down global processes of modernization of the last 500 years. It argues that in seven successive waves of modernization, the processes of democratization and communication have acted, through new technological, economic, political, and cultural formations, to integrate heterogeneous societies around the goals of security, liberty, equality, and community. However, tensions among these goals combined with the contradictions of social hierarchies of caste, class, status, race, religion, ethnicity, gender, and generation, have presented roadblocks to democracy. In order to be actualized, therefore, democratic values have to be contexualized into specific historical and cultural traditions. In the long haul, the broadening and deepening of the public sphere of discourse through interactive, oral or mediated communication, can overcome the obstacles to democratization. Governance by discussion is therefore the *sine qua non* of democratic development.

Democratization of politics and culture has been often too narrowly conceived in terms of the Western historical experiences, most notably in the rise of Athenian direct democracy in the fifth century BC, the development of the Roman legal system, and the cataclysms of the liberal democratic revolutions of the seventeenth to nineteenth centuries. But the roots of democracy must be sought, first and foremost, in the religious traditions that have asserted the equality of all men and women in the sight of God. Although the secular ideologies of progress (liberalism, nationalism, and Marxism) have mobilized the democratic forces in modern history, accompanied sometimes by the 'tyrannies of majorities' (De Tocqueville, 1956), the spiritual traditions of equality and civility have often (but not always) tried to temper the excesses. This chapter offers a global perspective on democratization that recognizes the contributions of all traditions, religious and secular, to the processes of this unfinished revolution. To provide an overview of a long historical process, it offers a conceptual framework that focuses on the interactions of democratization with modernization and communication.

The three processes are profoundly related in shaping the political, economic, and cultural domains of modern life. The question of which domain (economic, political, or cultural) is the primary mover of history has been debated among social scientists and historians *ad infinitum*. It is argued here that while the processes of modernization are mostly top-down, those of democratization are primarily bottom-up, while communication processes often mediate among and sometimes integrate the competing democratic values. However, a fuller understanding of this dialectic causality should be sought *not* in grand theories but in studies of historically specific times and places. Modernization is here viewed as the progressive application of rational, scientific, and technological knowledge to the processes of social production and organization. As reflected in the French revolutionary slogans of *Liberté, Égalité, Fraternité*, democratization may be considered, in addition to the security of life and limb, as the progressive realization of the threefold values of human liberty, equality, and community. But as Sir Ernest Barker once remarked, democracy is government by discussion. It is in the processes of public communication on public policies that democratic institutions thrive.

The narrowing of the public sphere leads to less politically literate citizens and more concentrated power in the hands of the few. The advent of each new communication technology in history has been accompanied by the rise of a new communication élite that masters the use of that particular technology and thus assumes a leading role in developing new mediating ideologies and institutions. The invention of writing, for example, was accompanied by the rise of the scribes and priesthoods who became the custodians of the new religious traditions; their holy writ was propagated by the new institutions of religious worship (temples, churches, mosques). Print technology augmented the power of the new scientific élite at the modern universities who established a new secular priesthood to challenge the authorities of the Church and the monarchies in the name of the natural, democratic rights of popular sovereignty.

The global context

Table 1 provides a schematic view of the unfolding of some 10,000 years of history. The table (1) suggests the combined and uneven development of the world in the three concurrent, contradictory, and overlapping processes of modernization, democratization, and communication; (2) reveals the collapses of time, space, and identities in an accelerating process of history; and (3) implies a deterritorialization of the world centres and peripheries of development in a process of creating a new, global cognitive élite linked through common education and access to information via the global electronic networks.

The metaphor of tsunamis, or tidal waves, caused by suboceanic earthquakes is employed here to suggest the enormity of the creative destructions that the forces of modernization and democratization entail. Tsunamis also suggest the invisibility of the forces they generate. However, a few caveats are in order. First, the table is constructed for heuristic purposes. The dates, in particular, should be taken seriously but not too seriously. They suggest some important but arguable historical watersheds. Second, although it may appear otherwise, the table does not present a stage theory of history. In fact, history is considered here as conjunctural and multilinear rather than progressive and linear. Third, the long period of history lumped together in the first tsunami may be considered as *anticipatory modernization* in that accumulation of science, technology, and capital in the ancient and medieval worlds provided the stock of knowledge from which the modern West heavily borrowed in order to pave the way to the Industrial Revolution. Fourth, the last tsunami is entirely hypothetical, taking us into the realm of science fiction.

Let us now review the table in its horizontal and vertical dimensions. Vertically, Table 1 can be read as a progressive collapse of time and space while, horizontally, it reveals a dual process at work towards homogenizing and differentiating human identities. The

Table 1: Seven modernizations and democratizations: Tidal waves or tsunamis in human history

Times	Spaces	Economies	Polities	Technologies	Ideologies	Communications
1) 8000 BC –1492 AD	Agrarian empires	Agrarian revolution	Feudal fiefdoms and multinational, bureaucratic empires; Democratization I: direct democracy and religious revolutions	Transmission of information: writing, ploughing, clay tablets, papyrus, roads, postal systems	Anticipatory modernization: rationalism vs. shamanism and religious dogmatism	Oral and written: shamans, soothsayers, poets, prophets, priests, temple
2) 1492–1648	City-states	Commercial revolution (aka mercantile capitalism)	Rise of city-states; Democratization II: Protestant Reformation and scientific revolution	Mechanization of information: print, compass, oceanfaring ships	Mercantilism vs. feudalism	Print: intellectuals, scientists, universities
3) 1648–1848	Nation-states	Industrial revolution (aka manufacturing capitalism)	Rise of nation-states and colonial empires; Democratization III: liberal democratic revolutions	Mass production of information newspapers, magazines, books, steam ships, trains	Liberal nationalism vs. monarchical absolutism	Élite media: publicists
4) 1848–1945	Industrial empires	Banking revolution (aka finance capitalism)	Multipolar world system; Democratization IV: social democratic and totalitarian revolutions	Electrification of information: telegraphy, telephony, photography, film, radio automobile, aircraft, paper money and banking	Imperialism vs. national liberation ideologies	Mass media: ideologues
5) 1945–1989	Planet	Managerial revolution (aka corporate capitalism vs. state capitalism/ communism)	Bipolar world system; Democratization V: national liberation revolutions	Digitalization of information: TV, computers, satellites, transborder data flows, electronic cash transfers, atomic energy; space probes	Globalism (capitalism vs. Communism) vs. nationalism	Big/small /medium technologues vs. communologues
6) 1989 –present	Cyberspace	Information revolution (aka technocratic capitalism)	Multipolar world system; Democratization VI: localist, ethnic, religious, feminist revolts	Integration of information: ISDN and multimedia, DSS, global networks (CNN, MTV, Internet)	Ecumenicalism vs. fundamentalism	Cybermedia: technologues, communologues vs. jestologues
7) Futures	Hyperspace	Space revolution (communitarian vs. totalitarian capitalism)	Cyborg planetary system; Democratization VII: underclass revolts against dehumanization	Totalization of information: timetravel, cyborgs, genetic engineering, spaceships and voyages	Totalism vs. communitarianism	Hypermedia: shamans, soothsayers, visuologues

collapses of time have revealed themselves in the accelerating rate of major technological breakthroughs (Toffler, 1970). Scientific and technological knowledge is an additive phenomenon feeding upon itself. It is not surprising, therefore, that it takes less and less time to move from one major technological breakthrough to another. Under the impact of rapid scientific and technological change, history is also visibly accelerating. It took, for instance, a century of struggle and two world wars for the edifice of modern European empires to collapse. But the Soviet Empire collapsed of its own accord in 1991 and within less than a decade. Similarly, the table shows that the succession of historical tsunamis is occurring at less frequent intervals as we move towards contemporary times.

The collapse of space is also visible from the table. Modernization as a process is incorporating progressively larger spaces within its domain of power, from localities to cities, nation-states, empires, the planet, the cyberspace of the global telecommunication networks, and the hyperspace of new dimensions of reality beyond the four of which we are aware (Kaku, 1994). In our own age of global communication, distances are being conquered by time. As in time travel, historical anachronisms also have become commonplace. Different historical epochs live side by side in the same spaces. As tsunamis collide on the surface of the ocean, so do civilizations and cultures. They borrow, steal, and adapt elements from their friends and enemies in order to advance their own cause. That is how an uneven world such as ours survives, adapts, moves forward and backward. An African village could juxtapose some of the signs of pre-modern, modern, and postmodern lives in a single space, encompassing the plough, the automobile, CocaCola, and Madonna's songs and semi-nude pictures. Art also imitates reality. The new post-modernist art juxtaposes times, spaces, and identities of different historical epochs in pastiches that shock and decentre our present sensibilities.

The fusion of identities is a more complex process but no less visible. The globalization of markets, communications, and cultures has led, from a pessimist view, to a CocaColonization of the world while, from an optimist view, it has resulted in global consciousness and citizenship. Whichever way we look at the phenomenon, it is certain that the awakening of progressively larger segments of humanity to historical consciousness and the increasing demands made by them to become subjects rather than objects of history has led to a differentiation of identities along religious, linguistic, ethnic, gender, and generational lines. The debate between the primordialists and constructivists on identity formations has served to demonstrate that humans construct their identities out of both past memories and present needs (Anderson, 1983). The collective memories are, in fact, politically negotiated to construct an identity that suits the perceived present and future needs of human communities. The differentiation and integration of identities is thus a dual process in history that has accelerated with higher levels of modernization, democratization, and communication. The individual and social constructions of identity have been thus facilitated by the increasing availability and projection of a rich repertoire of imaginaries from the past into the future.

These three historical processes have, however, given rise to enormous material and cultural disparities in the world, leading to a variety of conceptual maps. The social and political constructions of the world around such concepts as Occident and Orient (encompassing Europe and Asia), East and West (i.e. the communist and capitalist worlds), First, Second, and Third Worlds (i.e. the capitalist, communist, and non-aligned countries), North and South (industrial and pre-industrial nations), or Centre and Periphery (developed and underdeveloped countries) may all be considered as attempts to describe, analyse, and sometimes legitimate the world disparities in particular historical contexts. The complexities of the world, however, do not easily lend themselves to any single conceptual scheme. To allow for the increasing orders of complexity, some authors (Galtung and Vincent 1992) have even proposed a Fourth World that includes the newly industrialized countries (NICs) of East Asia (Japan, China, South Korea, Hong Kong, Taiwan, and the ASEAN). One could add to that a Fifth World (the least developed countries), and so on, in order to account for the increasing differentiation among the world nation-states.

The main problem with all of these conceptual maps is that they are basically territorial in conception. Global modernization and communication are, however, fast deterritorializing and informatizing the world. Some parts of the industrial world centres (notably the urban ghettos) are fast becoming peripherized (witness the rise of the unemployed and unemployable underclass in advanced industrial societies), while certain parts of the pre-industrial peripheries and industrializing semi-peripheries are achieving the high standards of

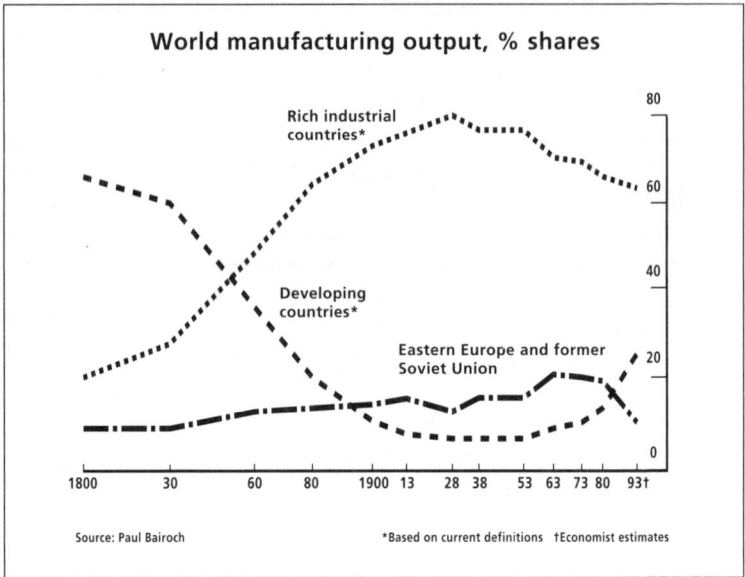

World manufacturing output, % shares

Figure 1: The cycles of history.   Source: Woodall, 1994: 9

living of the centres (e.g. Quandong Province in China, the city-states of Hong Kong and Singapore, the urban centres of the NICs). As the engines of the global economy, the transnational corporations (TNCs) have little regard for territorial boundaries. They go wherever lower wages, rents, taxes, and government regulation promise higher profits. The movements of capital and labour across the globe are further facilitated by the transportation, telecommunication, and tourism (TTT) technologies that have made central and global strategic planning possible when and if it is combined with decentralized operations by the corporate subsidiaries.

As the *Economist*'s Survey of the World Economy projects (Woodall, 1994), there is a possibility that sometime in the next century, the pendulum of history might swing back to Asia (see Figure 1). Around 1500, the adoption of the Chinese compass and the construction of oceanworthy ships by the Europeans led to the discovery of ocean routes to the Orient, opening the way for trade with and colonization of Asia, Africa, and Americas. The replacement of such land routes as the Silk and Spice Roads by the ocean routes weakened the Asians and strengthened the Europeans. By the mid-nineteenth century the developing countries lost their

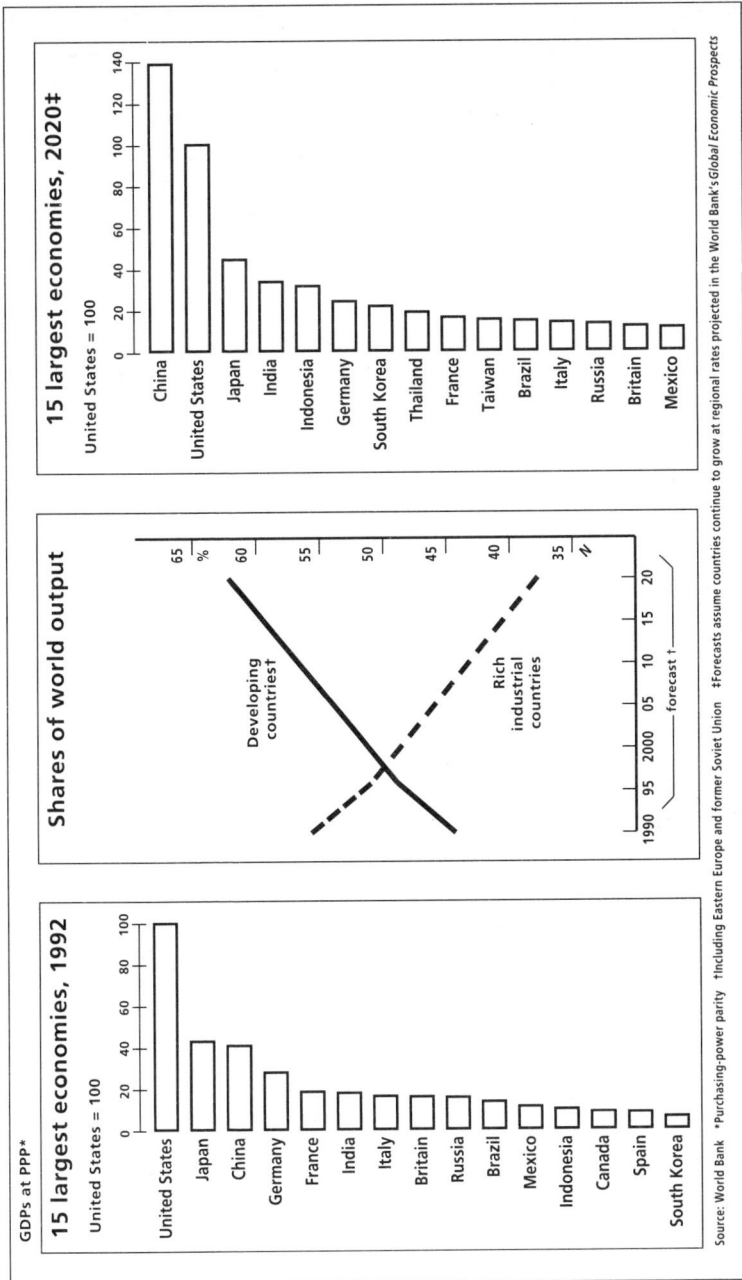

GDPs at PPP*

15 largest economies, 1992
United States = 100

Shares of world output

15 largest economies, 2020‡
United States = 100

Source: World Bank   *Purchasing-power parity   †Including Eastern Europe and former Soviet Union   ‡Forecasts assume countries continue to grow at regional rates projected in the World Bank's *Global Economic Prospects*

Figure 2: 2020 vision.   Source: Woodall, 1994: 4.

leading position in manufacturing to the developed countries. By technological leap-frogging and economic miracles, however, the Asians are now catching up. If the present rates of economic growth continue without any political mishaps, Asia may reach and possibly surpass the West by the year 2020 with China leading as the world's largest economy (see Figure 2).

Centres, peripheries, and semi-peripheries, therefore, can no longer be seen as neatly and permanently defined spatial categories; they should be conceived of as de-territorialized and informatized networks. The social status of individuals can no longer be assessed by their spatial location. An individual can live in New York and yet be worse off than another individual living in Bombay. She or he should be considered as part of a periphery in Harlem rather than of the core at Wall Street. The critical question in the emerging global hierarchy is whether an individual is informationally linked with the techno-structures of government, corporate, and related decision making institutions. Levels of wealth, income, education, and network connections clearly correlate positively with these informational and communication links. The chief actors in this global hierarchy are those who stand at the apex of the information pyramid. That is why the information rich are often the same as those actively engaged in the processes of global decision-making, while the information poor are subjected to the consequences of those decisions.

Seven modernizations

Table 1 provides a schematic outline of seven interlocking phases in modernization as experienced by the advanced industrial world. In this adaptive process, the less developed countries (LDCs) are currently emulating and leap-frogging the more developed countries (MDCs). Capitalism started with the spatial integration of the modern mercantile cities, moving successively on to the nation-state, imperial, global, and increasingly planetary and cyberspatial systems. If thus viewed, the processes of modernization can no longer be seen as a uniquely European phenomenon. Despite its dramatic results, modernization also cannot be considered as an abrupt change in the course of history. History crawls more than it leaps; it is evolving rather than making discrete departures from the past. Modernization should be more properly viewed as an accumulative process of accelerating change achieved by successive applications of scientific and technological knowledge. The pace of that change has historically gained an exponential momentum. Given the elements of continuity and change in this process, however, 'punctuated evolution' rather than 'revolution' is a more appropriate metaphor.

Democratization I

In each successive tsunami, the table draws attention to the times, spaces, economies, polities, technologies, ideologies, and communication institutions and élites of modernization. Sometime around 8000 BC, a transition from hunting and gathering to agrarian societies must have occurred (Frankfort *et al.*, 1963). That transition clearly required a higher order of complexity and organization than that of the earlier era. To tame the land, the domestication of animals and invention of agricultural tools were necessary. To assert the rights of individual or collective ownership to land and its fruits, accounting systems were required. Writing and numerology assisted in this process. But to record and transmit such information, clay tablets and papyrus came into play. Roads and postal systems were built to facilitate market exchange and central government control. But agrarian societies could not possibly survive without a rudimentary science that made technological innovation possible, however slowly. Transmission of information through writing and libraries in such places as Alexandria in ancient Egypt and Gundishapur in ancient Iran, became the indispensable tools of knowledge storage and retrieval for successive generations. Despite the enormous diversity of agrarian societies, for the sake of simplicity, we may consider this long wave of human evolution as anticipatory modernization lasting some 9,500 years.

The development of the democratic ideas of freedom, equality, and community during this long period is somewhat haphazard, but the universalist and egalitarian religions (notably Buddhism, Christianity, and Islam) clearly paved the way. The rejection of the traditional tribal, caste, or racial hierarchies have been at the heart of these religious traditions. In Buddhism, the possibility of spiritual enlightenment (nirvana or inner freedom from self-desire and its sufferings) is held up to all humanity without exception. In Christianity, Jesus offered his love and forgiveness to all, but considered salvation a privilege especially available to those who accepted his path and renounced the riches of this world. In Islam, all humanity stands as equal in the sight of God, but those professing Islam will be especially granted the peace of God.

The contributions of the secular creeds to democratization during this period is somewhat problematic. Athenian democracy in the age of Pericles provided a form of direct democracy by allowing citizens to participate in the affairs of the state by discussing and voting on public issues, but citizenship was denied to women and slaves. The Roman Empire similarly limited its democratic privileges to the citizens of Rome while women and slaves were excluded. Other imperial systems in China, India, Persia, and Russia were largely based on the divine right of kings and absolute powers. Such famous legal documents as the Code of Hammurabi (1750 BC) and Magna Carta (AD 1215) may be considered as the precursors to our own contemporary Bill of Rights, but their rights and privileges were limited to a small aristocracy.

The role of communication and communicators in the earliest processes of democratization is pivotal. The Agora in Greece, the Roman Senate, and the Ka'aba in Mecca may be considered as the precursors of our own modern public sphere. In such forums, the shamans, soothsayers, poets, philosophers, orators, and prophets could address a large audience publicly on matters of public importance. However, the transition from the oral to written traditions created a new cognitive and communication élite that replaced the orators. The rise of writing gave birth to the holy writ, and a new class of custodians who established their own religious institutions and priestly hierarchies. The emergence of religious institutions as distinct from the state institutions provided a basis for resistance against the arbitrary powers of rulers. To the extent that these institutions enjoyed autonomy in sources of financing, drawn from the guilds and merchants, they also exercised some check and balance *vis-à-vis* the state.

The seeds of commercialism and a kind of primitive capitalism were sown on the world's first global economy – the Silk and Spice Routes. Along these routes, in the ancient cities of Beijing, Xian, Kashqar, Samarkand, Bukhara, Neishapur, Rey, Isfahan, Baghdad, Mecca, Medina, Bombay, Damascus, Allepo, Constantinople, Athens, and Rome, urban civilizations developed that housed industrial arts and crafts, observatories, universities, and libraries. Under the weight of rapacious agrarian empires, however, they constantly waxed and waned. The rise and fall of these multinational, agrarian, imperial systems closely correlated with the rise of fall of the cities. The absence of autonomous municipalities and independent merchant classes to support their continuous prosperity often led to cycles of urban rise and ruin. Imperial systems provided for expanded security and trade, but their decline often caused political fragmentation and breakdown of trade routes.

Democratization II

The genius of modern Europe was in its ability to break through this vicious cycle. A second tidal wave of modernization set the Western European scene by the sixteenth century. The important historical watersheds in this tsunami were, in 1492, the final expulsion of the Muslims from Granada in Spain while Columbus was stumbling into a New World in the same year. Reformation and Renaissance followed shortly thereafter. Beginning with Martin Luther's posting of ninety-five theses on the church door at Wittenberg, in 1517, the Protestant revolt against the Catholic Church signalled a religious war that continued to the end of the Thirty-year War and the Peace of Westphalia in 1648. The Reformation took many different forms in Europe, but its Protestant ethics of frugality, hard work, savings, and investment clearly ushered in a new spirit of capitalism (Weber 1958; Tawney 1984). Similarly, the Renaissance lasting from the fourteenth to the mid-seventeenth centuries introduced a new spirit of humanism in arts and letters, scientific exploration, and commerce.

The combined effects of Reformation and Renaissance were threefold – privatization, nationalization, and secularization. First, through the Protestant challenge to Papal authority and Catholic hierarchy, religious faith and morality in Christendom became increasingly privatized, i.e. a matter between God and the individual's conscience rather than an issue in the hands of the intermediary, priestly class. Individual responsibility and freedom of conscience thus gave a further impetus to the processes of individuation that modernity entails. The distinction between the private and public spheres of discourse also was in part a consequence of this privatization. Second, as the Protestant revolution spread throughout Europe and the Americas, it branched into a complex variety of national and sub-national churches that employed the vernacular languages instead of Latin, following particular national rather than Catholic (i.e. universal) traditions. The Lutheran Church in Germany and the Church of England were the first national churches to declare their independence from Rome, but others soon followed. The national Protestant churches owed their growth in no small measure to the rise of print technologies that made the translation and publication of the Bible in the vernacular, European languages possible. Third, the Renaissance, combined with the scientific revolution of the seventeenth century, led to the rise of a secular culture that, when combined with the privatization of faith and nationalization of religious institutions, provided a powerful ideological force towards modernization. That, in turn, rationalized the religious institutions away from their magical foundations (miracles, superstitions) towards more worldly concerns.

However, the second tsunami of modernization clearly owed its material force to the rise of the modern mercantile cities in which the seeds of modern capitalism were sown by the monetization and commercialization of the European economies. Urbanization during the Middle Ages was a very slow process but it created large cities such as Venice, Florence, Amsterdam, Lisbon, Barcelona, London, and Paris serving as the centres of a new urbanism. Although large cities were characteristic of the ancient Asian and African civilizations as well, the new European cities were markedly different in that they enjoyed some measure of municipal autonomy. Whereas Rome, Constantinople, Baghdad, Cairo, Isfahan, Delhi, Beijing and Timbuktu served as the urban centres of multinational, agrarian empires, the new European cities were breaking from the agrarian past into a new era of mercantile capitalism. The Crusades during the eleventh to thirteenth centuries, technological breakthroughs in ocean transportation, the discovery of the New World in the late fifteenth century, the flow of newly discovered gold from the New World to the Old World, the scientific revolution of the seventeenth century, the rise of manufacturing in the new towns, the Enlightenment movement in the eighteenth century, and the political revolutions of the eighteenth and nineteenth centuries – each contributed to a transfer of economic and political power from the feudal lords and manors to a new class of merchants in the new mercantile towns and cities under a new capitalist order. Heilbroner (1962: 48) has aptly summarized the process:

> When the travelling merchants stopped, they naturally chose the protected site of a local castle or burg. And we find growing up around the walls of advantageously situated castles – in the *focus burgis*, whence *faubourg*, the French word for 'suburb' – more or less permanent trading places, which in

turn became the inner core of small towns. Nestled close to the castle wall for protection, the new burgs were still not 'of' the manor. The inhabitants of the burg – the burgesses, burghers, bourgeois – had at best an anomalous and insecure relation to the manorial world within. As we have seen, there was no way of applying the time-hallowed rule of 'ancient customs' in adjudicating their disputes, since there *were* no ancient customs in the commercial quarters. Neither were there clear-cut rules for their taxation or for the particular degrees of fealty they owed their local masters. Worse yet, some of the growing towns began to surround themselves with walls. By the twelfth century, the commercial burg of Bruges, for example, had already swallowed up the old fortress like a pearl around a grain of sand.

The commercial revolution and its cultural orchestration by the Reformation, Renaissance, and the scientific revolution of the seventeenth century thus contributed to the development of a modern, democratic cultural system that put the individual at the centre of social responsibility. The dual challenge to the authorities of the Church and monarchy on behalf of individual rights and responsibilities of citizenship received its ideological impetus from a new communication élite, the intellectuals, who were situated in the modern, secular universities while riding on the wave of the new technology of print. Print provided not only direct access to the holy writ, pluralizing its interpretations, it also allowed the new secular, scientific, and philosophical ideas to spread among an increasing number of the literate population (Eisenstein, 1979).

Democratization III
Whereas in the processes of anticipatory modernization it was often the countryside that controlled the cities in the form of land-owning feudal aristocracies or tribal conquests, the growth of modern cities signalled a reversal of that role. In fact, the new cities' superior technological and organizational abilities for accumulation of capital led, in turn, to a Third Modernization – the rise of the territorial nation-state system. It is customary to consider the Peace of Westphalia in 1648 as the beginning of the modern nation-state system. As an island 'nation', England led the way in this process demonstrating, in effect, the advantages of the nation-state as a new form of political and spatial organization. The ideology of the new political-spatial entity was nationalism, its religion was embodied in a national church such as the Church of England and other Protestant

churches declaring their independence from the Papacy; its language was no longer Latin, but instead the vernacular languages such as English, French, German, and Italian; its ethos became the rationalized Protestant ethics which, as Max Weber (1958) has argued, emboldened 'the spirit of capitalism' towards hard work, frugality, saving, and investment. Although the Catholic countries (France, Italy, and Spain) did not break away from Rome, they too came increasingly under the impact of a nationalism that fostered the secular spirits of rationality, science, and technology. In comparison to the imperial or city states of the past, the nation-state system clearly supplied a far more homogeneous population, greater defensible natural boundaries, economies of scale, and superior social and political opportunities for mobilizing the population on national missions of consolidation and expansion.

The United States was the 'first nation' (Lipset, 1963) to come about through a written constitution, demonstrating how the imagery of nationhood could be realized more through solidarity of democratic ideas than ties of blood and race. Following the American example, the construction of the nation-state system reached its fruition in Europe during the nineteenthth century by the unifications of the German and Italian states into two new nation-state systems. The system received further impetus following the First World War by the breakdown of the Ottoman, Austro-Hungarian, and Russian empires and the emergence of new nation-states in the Balkans and the Arab world. The Second World War extended the system to Asia, Africa and Latin America through the breakdown of the European imperial systems. In 1991, the breakdown of the world's last empire, the Soviet Union, led to a further proliferation of the nation-states.

The Third Democratization was mainly achieved through the liberal democratic revolutions of England (1688), the United States (1776), and France (1789). The ideas of liberalism and nationalism combined to provide the basis both for free trade and protectionism. The tension between the two imperatives of capitalist growth, at the domestic and international levels, became the main source of international conflict in the next century of imperialist rivalries. In the meantime, however, the Enlightenment *philosophes* in France, the liberal political economists in England, and the Federalist publicists in the United States had sown the seeds of a universal doctrine of liberty and equality that could and *did* later undermine the empires. Collectively known as the contractualist theorists (notably John Locke, Thomas

Hobbes, Jean Jacques Rousseau, and the American Federalists, Hamilton, Jay, and Madison), they posited a doctrine of natural rights that viewed individuals in the state of nature entering into a contract to form a government that would protect their life, liberty, and property (or pursuit of happiness, in the Jeffersonian formulation). This was a radical departure from the divine right of kings towards a doctrine of popular sovereignty that became the cornerstone of most democratic constitutions. The mass production of information (newspapers, magazines, books) and mass transportation (steamships and trains) rapidly spread the new ideas around the world.

Democratization IV

As old empires were collapsing, however, a new, industrial imperialism was being born in the nineteenth century that may be considered as the Fourth Modernization. As Hobson (1902) and Lenin (1917) first recognized, the new imperialism was materially different from all empires of the past. It was built on the foundations of the new capitalist order, driven by the search for new sources of raw materials, cheap labour, and consumer markets. Flag followed trade rather than the other way around. Military conquest served only as the last resort for opening up new territories for capitalist exploitation. The shifts from small-scale, entrepreneurial capitalism to large-scale, corporate capitalism, and from manufacturing domination to the supremacy of finance capital, all happening in the second half of the nineteenth century (Trachtenberg, 1982), made the growth of the new empires both possible and, from the capitalist point of view, desirable. Although Pax Britannica ruled the world between the end of the Napoleonic Wars in 1814 and the outbreak of the First World War in 1914, the foundations for a multi-polar world were formed by the rise of the United States, Japan, Germany, and Russia as rival super-powers. The two world wars of the twentieth century could be, in fact, viewed as struggles in the imperialist game between the old players (England, France, and Russia) and the new players (Japan, Germany, Italy, and the United States).

Democratization took a new twist in this phase by developments against the centre of the political spectrum led by the leftist and rightist forces. The year 1848 was, ironically, both the year of liberal democratic revolutions in Western Europe as well as that of the publication of *The Communist Manifesto*. In that brilliant tract, Marx and Engels predicted a proletarian revolution, first in Germany and subsequently elsewhere, while the logic of their own argument called for a global triumph of capitalism before socialism could be established. Their revolutionary enthusiasm had thus blinded them to their own more sober logic. Nevertheless, the class contradictions of liberal capitalism had already created two types of grave-diggers for liberal democracy, i.e. a revolutionary working class and a disenchanted lower middle class, both led by a new type of communication élite, the *ideologues*. Electrification of information through telegraphy and telephony had facilitated the communication systems of not only the empires but also those of the oppositions. Mass circulation newspapers, radio broadcasting, photography, and film also added new arsenals to mass movements in Europe and Asia challenging the authority of the imperial systems. Lenin, as the chief ideologue of the left, revised the Marxist theories to argue that the socialist revolution need not wait for the worldwide triumph of capitalism; instead it could strike at the weakest links of the capitalist-imperialist chain in Russia and the colonial world. Hitler, as the chief ideologue of the right, revised the liberal theories to argue for the hegemony of capital within a national, rather than international, socialist framework. The populism of both ideologies drew millions of the working and lower middle classes within the vortex of movements that, despite their democratic appeals to egalitarianism, ultimately proved to be totalitarian in nature. In Germany, Italy, Japan, and Spain, it united the capitalist class with the lower middle classes against the liberal and communist movements. In the Soviet Union, it led to the rise of a bureaucratic state capitalist cum communist regime.

The rise of totalitarian regimes of the right and the left proved that the transition to the modern, industrial world need not take place through a liberal democratic regime. In fact, the late-comers to industrialization are often faced with a pathology of transition that calls for hyper-modernization by means of central planning and regimentation. The totalitarian temptation may be thus considered as inherent in the processes of modernization, particularly when class contradictions destroy or seriously threaten the ruling élites. In the former case, it often leads to a revolutionary movement, supported by the working class and peasantry, that has little regard for civil liberties. In the latter case, it usually brings about an alliance between the ruling élites and the mobilized lower-middle classes united in their struggles against liberalism and communism.

Democratization V

The Fifth Modernization may be considered to have begun in 1945, with the rise of globalism at the conclusion of the Second World War. Two factors were paramount at this historical juncture – the collapse of the European empires and the pre-eminence of the United States. Fortified by the establishment of the United Nations on the basis of the principle of national self-determination, the national liberation movements in the colonial world destroyed the old and exhausted European empires that could no longer endure as viable forms of political-spatial organization. As a former colony itself and a new superpower, the United States was well poised to lead the way towards a new principle of globalism. The Bretton Woods Agreement provided the economic basis for such a global capitalist system by creating a World Bank to channel investments from the more to the less developed countries, an International Monetary Fund (IMF) to manage the international currency exchange convertibility, and the General Agreement on Tariffs and Trade (GATT) to encourage world trade by reducing the tariff and non-tariff barriers.

The United Nations also was to provide the political basis for a global collective security system guaranteed by the five permanent members of the Security Council. However, the unanimity needed for such a system broke down with the onset of the Cold War in 1947. The wars in Korea and Vietnam further undermined the UN collective security system by pitting the First World (i.e. the capitalist countries) against the Second World (i.e. the socialist countries), largely in alliance with the revolutionary parts of the Third World. Nevertheless, globalism continued its irresistible growth through 'the global reach' (Barnet and Muller 1974) of the transnational corporations (TNCs). In their search for new sources of raw material, consumer markets, investment opportunities, low taxes, low rents, low wages, and low government control, the TNCs devised global strategies that ensured them centralized control but spatial and managerial dispersion. Globalism also has been further strengthened by the reach of global advertising and the culture of mass consumption that it fosters. The seductions of the 'soft power' (Nye 1990) of cultural appeal have proved nearly as powerful as the 'hard power' of military might and economic gravity.

Globalism, however, appeared in two distinct versions – liberal capitalism and Soviet communism – each covertly or overtly opposing the nascent nationalism of the post-colonial world. During the Cold

War, the Enlightenment idea of progress was presented to the world in the the form of two options – development by the market rules or by central planning. The rivalries of the two superpowers for the hearts and minds of the rest of the world was a total war that employed both the cool methods of ideological struggle as well as the hot instruments of naked violence. In places such as Korea, Vietnam, Cambodia, Iran, Congo, Angola, Somalia, Chile, Nicaragua, Cuba, and Afghanistan, it erupted into military confrontations or CIA subversions that assumed dramatic importance. Both camps employed all of the media arsenal at their disposal – print, film, radio, television, satellites, and computers. Radio Free Europe, Radio Liberty, the United States Information Agency (USIA), and American friendship societies were government agents. They were supplemented by such commercial media as *Time*, *Life*, Hollywood films, global advertising, and direct broadcast satellites (DBS). Although less powerful and pervasive, the Soviets and the Chinese communist regimes also attempted to compete in all of these fields, by equal means or by jamming the enemy radio frequencies. In Eastern Europe and the Caribbean, radio jamming sometimes assumed wartime proportions.

The Big Media (BBC, Voice of America, CNN, MTV, Hollywood films, global newspapers such as the *Wall Street Journal* and the *International Herald Tribune*, and global news magazines such as *Time*, *Newsweek*, and the *Economist*) have served as the ideological vehicles of the First World. Their equivalents in the Second World have been Radio Moscow, Radio Beijing, *New Times*, and *Beijing Review*. On both sides of the ideological camp, however, a new and distinctly different communication élite was managing the huge technostructures of global communication. Standing apart from earlier print generations of *publicists* and *ideologues*, and riding high on the technological wave of digitalization of information, the new communication élite may be called *technologues*. The convergence of voice, data, and images into a single stream of electronic, digital signals known as ISDN, the Integrated System Digital Network, has been blurring the boundaries among media technologies, institutions, and professions. For the technologues, knowledge of the emerging technologies has to combine with the managerial skills for running large and complex technocracies. In the advanced industrial societies, the locus of power seems to have decisively shifted from ideas and ideologues to techniques and technologues.

In the Third World, however, where the level of technological development has lagged behind, other types of communication élite have been in formation. The intellectuals and ideologues, associated with the print and broadcasting media, are situated in the emerging institutions of universities, mass media, and mass political parties. However, the broadening and deepening of mass movements into the more traditional sectors of the population has led to the rise of a more recent communication élite that may be called *communologues*. In contrast to the intellectuals and ideologues who speak in the universal grammar of the Enlightenment project (namely liberalism and Marxism), the communologues converse in the vernacular of the indigenous religious, ethnic, and nationalist discourses. In contrast to the intellectual-ideologues of an earlier generation of Third World leaders (such as Nehru, Mosaddeq, Nasser, Sukarno, Mao, Chou Enlai, Nkrumah, and Nyerere), the communologue leaders such as Gandhi in India, Khomeini in Iran, Shaikh Fazlallah in Lebanon, and the Theology of Liberation priests in Latin America, come from a religious perspective, rejecting the secular views of progress while reconstructing their own indigenous Hindu, Islamic, or Christian traditions. When combined with mass movements, the small media of mimeographing and copying machines, audio- and video-cassette recorders, transistor radios and, increasingly, desktop publishing, have given the new communication élite a power to challenge the authority of the ruling governments and their big media.

The rise of political religion is not, however, limited to the Third World. The Electronic Church in the United States and its offshoots in Latin America, the re-emergence of the Russian Orthodox Church in the former Soviet Union, the rise of ethno-religious conflicts in the former Yugoslavia and the Republics of Armenia and Azerbaijan, all seem to suggest that religion is once again presenting a powerful source of political energy for a variety of social movements from the right as well as the left of the political spectrum. In fact, the terms 'right' and 'left', originating in secular French revolutionary ideas and practices, have lost some of their analytical power in understanding neotraditionalist and postmodern politics.

The rejection of the grand metanarratives of evolutionary progress is a hallmark of postmodernism. The emergence of postmodern architecture and city planning in reaction to the preponderance of the rationalist, glass cages of modernist architecture and urban design may be viewed as part of the transition from the Fifth to the Sixth

Modernization (K. Tehranian, 1994). The universalism of the modernist, international style was challenged by the eclecticism and pastiche quality of postmodern architecture. The utopian cities of the modern world have been similarly giving way to the 'collage cities' of the postmodern era (Rowe and Koetter, 1992). Rejecting the grand utopian visions of total planning and total design, the postmodernists are calling for an eclectic city of many faces and neighbourhoods that can accommodate a whole range of utopias in miniature.

Postmodernist cultural trends have similarly challenged the metanarratives of modernist culture. Exposure to the diversity of global cultures has, however, challenged the foundations of globalism as an ideology of modernity. It has decentred its universalist claims of the Enlightenment project and created a new sense of cultural relativism. It also has juxtaposed localist cultures and practices alongside the global artifacts of skyscrapers, McDonalds, CocaCola, and Hilton Hotels. Thus, the tension between the global and the local has been an incipient political and cultural feature of the Fifth Modernization. The unresolved contradictions have erupted into the open with greater gusto following the end of the Cold War and a balance of terror between the two superpowers. The military strategy of mutually assured destruction (whose appropriate acronym is MAD) ensured an international equilibrium that was disturbed only during crises and brinksmanships such as the Missile Crisis of 1962.

Democratization VI
The Sixth Modernization may be dated from 1989 when at the meeting of Presidents Bush and Gorbachev in Malta, the Cold War came to an end. From Yalta to Malta, the world witnessed an uneasy equilibrium between the two superpowers, each armed with weapons of mass destruction sufficient to annihilate human civilization not once but several times. With the demise of the Soviet Union and the more gradual slippage of China into the vortex of world capitalism, the capitalist version of globalism seems more irresistible in the 1990s than ever before. The passage of GATT in the United States Congress, in late 1994, and its impending ratification by some 123 countries, signals another new and significant departure towards globalism. Known for years as the Global Agreement to Talk and Talk, the new GATT will be enforced by a newly established World Trade Organization (WTO), including panels of three judges from

countries other than the disputants who would rule on trade disputes. WTO was part of the Bretton Woods Agreement in 1944, including the World Bank (IBRD) and the International Monetary Fund (IMF), all designed to manage the post-war world economic system. However, the US Congress rejected WTO as an infringement upon US national sovereignty. In its place, in seven successive rounds of talks, GATT became the channel for the negotiation of world trade issues. The passage of the most recent GATT treaty is expected to lower tariffs by some 38 per cent, resulting in a gain of some $744 billion for the participating 123 countries, and hastening the processes of global modernization.

At the same time, however, the global forces unleashed by GATT are likely to undermine the social fabric of developed and developing countries. The domestic cleavages resulting from rapid technological change combined with huge transfers of capital and large-scale dislocations of labour may prove too powerful for some regimes. The existence of some 23 million refugees around the world is already straining the capacity of some countries to sustain the burden of legal and illegal immigrants. In reaction, such measures as Proposition 187 passed in California in November 1994, are bound to fuel international ill feelings (in this case between the United States and Mexico). The phenomenon of 'ethnic cleansing' in Bosnia, Tajikistan, Rwanda, Somalia, Palestine, Kashmir, the Muslim-Hindu conflicts in India, and Armenian–Azerbaijani conflicts in the Caucasus, suggest an alarming rise in tribal, ethnic, and religious hatreds. Even more alarming is the fact that the great powers are not willing to intervene in such situations unless and until their vital economic or political interests are at stake. Compare the active and massive interventions in the Persian Gulf crises of 1990–91 and 1994 to the lukewarm and failed interventions in Somalia, Rwanda, Tajikistan, and Bosnia. The rise of isolationism, appeasement of the aggressors, and domestic indifference or hostility to the plight of the minorities are reminiscent of what happened in the interwar period. Those developments led to the rise of fascism in Europe resulting in the Holocaust. The failure of the international community to stop these aggressions has historically proved to be an open invitation to other expansionist aggressors.

Clearly the world established after the Cold War is not an orderly and peaceful place. The new world order based on the rule of international law, promised by President Bush in 1990, was stillborn. The old world order of capitalist hegemony is continuing unabated but with a declining liberal inclination and a rising conservatism. However, several sources of resistance to the global hegemony of capital deserve notice. Nationalism is still a force to be reckoned with. It continues to provide the primary and irreducible principle of spatial-political organization in the world for nearly 200 recognized states, and holds the promise for some 5,000 stateless nationalities in search of political and spatial recognition. Regionalism provides another source of resistance to globalism, with which it may or may not co-exist. Regional organizations such as the European Union (EU), the Association of Southeast Asian Nations (ASEAN), the North American Free Trade Area (NAFTA), and the Asia-Pacific Economic Co-operation (APEC) present actual and potential trading blocs which could turn into fortresses if and when the global system breaks down through economic or political crises. Localism within each large nation-state such as the United States, Russia, India, China, or Brazil also presents a source of resistance to the global hegemony of a capitalist order. Last but not least, religious revivalism, in the form of a variety of fundamentalist movements, could disrupt the rule of capital in important parts of the world. The fundamental problem is that while capitalism has unleashed immense productive possibilities in the modern world, it has also created great and growing disparities among and within nations. The rule of capital has therefore led to prosperity for the privileged sectors of the population, some trickling down, but not enough to guarantee social peace and solidarity. The global capitalist system, though ever growing and gaining, continues to be vulnerable – technologically, economically, politically.

The vulnerability of global capital is best demonstrated in the dual effects of its technologies in transportation and telecommunication. The new spaces of modernization are the airlanes of modern jet transportation and the de-territorialized cyberspace of modern telecommunication. International terrorism on the global airlanes and urban spaces, and international subversion through the cyberspaces of the global telecommunication networks provide some evidence of this vulnerability. Both of these technologies have increasingly blurred the conventional distinctions between the global and local. As the events in the Persian Gulf, Somalia, and Bosnia have demonstrated in recent years, the global conflicts are often localized, and the local conflicts are as often globalized. Similarly, domination and resistance are simultaneously assuming global and local dimensions. The Iranian Revolution, the Iran–Iraq War, the Persian Gulf War, and the bomb explosion at the New York World Trade Center cannot be understood except by correlating the global with the local issues of the conflict.

The world system in this wave of modernization appears to be both unipolar and fragmented. With the decline of Russia, the inability of the European Union (EU) to speak in one voice, and the continuing reluctance of Japan and China to assume world responsibilities beyond their own region, the United States has assumed the role of the sole balancer of power. This is a role that Britain played during the nineteenth century. When the United States acts with determination, as in the Gulf War, others follow. When it vacillates, as in Bosnia, the rest also waver and jockey for position. In the meantime, however, regional groupings such as the EU, CIS, NAFTA and ASEAN are establishing powerful and competing economic and political blocs. At a summit meeting in December 1994, the heads of thirty-four states in the Americas decided to establish a Pan-American free-trade area by the year 2005. As NAFTA or its Pan-American successor enlarge into the whole of Latin America, it would become the largest free-trade zone in the world, encompassing some 850 million people and $13 trillion in combined purchasing power.

The forces of globalization and regionalization are thus homogenizing the markets and styles of life at an accelerating rate. At the same time, however, the rapid diffusion and miniaturization of communication technologies are providing the vehicles for the expression of nationalist and localist voices that are threatened with obliteration. As Robertson (1994) and others have argued, *glocalization* seems to be a dominant feature of the postmodern world. Indeed, the processes of globalization can be viewed at the same time as processes of localization. The global market is adapting to the local conditions while it employs them to gain competitive advantage. The global communication network is globalizing local issues (e.g. Bosnia, Tajikistan, Kurdistan) at the same pace that it localizes global issues such as the environment, human rights, and population control. Global forces valorize local traits and faces in the dissemination of such consumer items as food, tourism, modelling, arts, and crafts. The top-down processes of globalization are thus working concurrently with the bottom-up processes of localization. Glocalization thus appears to be the wave of the future.

Democratization in the Sixth Modernization is also entering a new, broadening and deepening phase. In the absence of bipolar Cold War rivalries, the localist forces of ethnic, religious, and tribal loyalties are re-surfacing to challenge the authority of the existing state systems. Fred Riggs (1994) has argued that the forces of ethnonationalism have come through three tsunamis. The first wave came with the rise of European nationalism in the eighteenth and nineteenth centuries followed by a second wave of national liberation movements in Africa, Asia, and Latin America in the twentieth century. The current, third wave of ethnonationalism is witnessing the rise of those ethnic and racial minorities (and sometime majorities as in South Africa) who have been repressed by the nation-states of two previous waves.

In this respect, the example of the former Soviet Union is rather telling. Ethnonationalism is apparent in most of the post-Soviet republics because in the name of proletarian solidarity, the Soviet ideology had suppressed nationalist and religious expressions for some seventy years. The Soviets were rather successful in state building, i.e. the development of military, civilian, and police structures as well as the educational, transportation, and communication infrastructures of the state. However, by design, they miserably failed in nation-building. Although Stalin's doctrine of national self-determination for the fifteen Soviet republics was theoretically the state policy, in practice, the borders were drawn in such a way that each republic contained significant numbers of ethnic and religious minorities within it. As immigrants from the European parts of the Soviet Union moved into the Asian and Caucasian republics, the multi-ethnic nature of the republics was further reinforced. The imperial policies of divide and rule also kept the regions and ethnicities fairly separate and often at odds. For a combination of the foregoing reasons, the dissolution of the communist regimes in the former Soviet Union and its Eastern European satellites has been accompanied by a resurgence of ethnic and religious conflicts.

During the Sixth Modernization and Democratization, therefore, two distinctly different types of ideologies and pathologies are simultaneously at work. These may be characterized as globalism vs. localism, and commodity fetishism vs. identity fetishism. The global market-place clearly favours the secular ideologies of progress that encourage an acquisitive society and competitive individuals. Commodity fetishism, i.e. a desperate struggle to acquire the material symbols of modernity, seems to be therefore an intrinsic pathology of the modern world. Material poverty in the age of modernity is no longer a condition that can be borne with dignity. In this respect, modernized urban poverty fundamentally differs from the rural poverty of the pre-modern world. The urban poor in the modern centres of industrial and financial power live in the constant company of the rich, spatially and symbolically. Through television signals they are exposed to the standards of living among the rich and the middle classes, while through their menial jobs, they occasionally come into contact with those whom they envy. Since the dominant culture holds up material success as a sign of superior moral standards, they also are incessantly reminded not only of their economic but also 'moral' failure. Modern poverty therefore induces ceaseless anxieties, feelings of shame and worthlessness, and frustrations that lead to regression and aggression. As the high rate of self-inflicted violence among black youth in the United States shows, the aggression is often directed against one's own community. But when it finds a legitimate cause, it can be directed against the outside world as well. The regression to an earlier stage of dependency often leads to identity fetishism, a pathology that through collective identities, loyalties, and actions breeds a sense of false security in an uncertain and threatening world. The mass hysteria and behaviour under the conditions of inter-national, revolutionary, civil, and gang warfare are symptomatic of such a pathology. Totalitarian ideologies such as fascism, com-munism, and fundamentalism thrive under such conditions.

Totalitarianism may thus be viewed as a pathology of transition. By uprooting the social fabric and traditions of civility, rapid technological and social change provides the breeding ground for the ceaseless anxieties of commodity and identity fetishisms. The only recourse appears to be more, not less democracy. The homogenizing tendencies of totalitarianism can be checked only through an acknowledgement of social and cultural diversity and the develop-ment of civil societies that reflect that diversity by the formation of voluntary associations and free-speech communities. The newly emerging, interactive technologies of communication seem to have a bias for democracy. But they can be also used for surveillance in an information-perfect society. The dual potentialities of the communication and information technologies can be best seen in electronic eavesdropping as well as electronic town meetings, satellite remote sensing as well as direct satellite broadcasts, computerized surveillance as well as the Internet's virtual communities.

The new cybermedia, characterized by interactivity and

convertibility, are giving rise to several different and often contradictory types of communications élites, namely *technologues, communologues,* and *jestologues*. The impact of computer technologies on every aspect of economic and social life has created a new class of *technologues*. But the diffusion of the small media of communication has boosted the power and influence of the traditional communication élites (the priests, the mullas, the monks, the community activists), i.e. the *communologues*, who can speak in the vernacular languages of common folk. The demystifying power of visual media (television, cable, and VCRs) seems to have led to a new and sceptical generation of communication élite that sees through the pretensions of the ideologues, technologues, and communologues. The new communication élite serves the same function as the jesters and clowns in the kings' courts. Hence, we may call them *jestologues*.

Jestologues mock the powers that be with a humour that is often tolerated, but they also risk their heads. As pioneers of a new culture of postmodernity, they are jestful, relativistic, antinarrative, despairing, ecstatic, playful, and self-mocking. The war between communologues and jestologues, between neotraditional modernity and postmodernity, was officially declared by Ayatollah Khomeini's death warrant on Salman Rushdie. While the Ayatollah and his followers have been committed to the sacred mission of realizing the Kingdom of God on earth, Salman Rushdie poses as the postmodern jester who mocks all sanctities. Most interpretations of the confrontation between the two camps have portrayed the Ayatollah as the traditional, religious bigot and Salman Rushdie as the modern, free-thinking intellectual. But the two figures and what they stand for in the contemporary world can be perhaps better understood if we view each in terms of some of the distinctions made between pre-modern, modern, and postmodern. The Ayatollah and his successors are a complex mix of pre-modernist and modernist Islamic leaders in their neotraditional, totalizing strategy of fusing the state and the mosque into a single theocratic regime (Tehranian, 1992). In contrast, Salman Rushdie is a postmodern critic in his deconstructionist strategy of mocking the traditional and modern sanctities. The postmodern strategy is to shock, to startle, and to decentre in order to dethrone the sacred and the naturalized. Its paramount medium is the musical video, which has developed a nearly universal and irresistible language in the global MTV channel. Its heroes are the deconstructionist anti-heroes (Beavis and Butthead), the new self-mocking shamans of electronic rock music (e.g. Sting or Bono), or the glittering stars of multiple identities and sexualities (e.g. Madonna and Michael Jackson). The conflict between the pre-modern, modern, and post-modern is thus part of the cultural landscape of an economically uneven, politically tribalized, and culturally schizoid, contemporary world.

Democratization VII

The contradictions of the Sixth Modernization and Democratization will be, no doubt, carried into the Seventh. The film *Blade Runner* has given us a cinematic glimpse of what the Seventh Modernization might look like in Los Angeles in the year 2019. As I am writing, the Hubble telescope is scanning the universe for other planets and life forms. Michio Kaku (1994), a leading theoretical physicist, argues that string theory has mathematically deduced the existence of ten dimensions of reality. As in *Flatlands*, the two-dimensional beings cannot conceive of the three-dimensionals except as fleeting sensations. As four-dimensional beings (three of space and one of time), the humans also cannot perceive the higher dimensional beings except as fleeting sensations. The spaces of the future to be conquered and understood are therefore the ten-dimensional *hyperspaces* of theoretical physics. By then, cyborgs will be an important part of the systems of production, distribution, and information. Democratization VII will be therefore characterized by a struggle between the economic-political-scientific élites of the future allied with their cyborgs against the masses of humanity whose numbers will swell into billions threatening them with periodic revolts. If we are to believe the current crop of science fictions, genetic engineering, time travel, and space voyages will be part of the routines of life sometime in the twenty-first century. But the struggle between good and evil, democracy and tyranny, shamans and jesters, communologues and jestologues will not cease.

Prospects for democracy

Modernization and democratization may be viewed as dialectical processes in which the requirements of economic accumulation and political participation are competing for the material and symbolic resources of power (Tehranian, 1990, ch. 9). In the early stages of primitive accumulation, the need for high levels of national savings and investment often leads to forced savings through low wages. In later stages of development, the demand for higher wages is often expressed by the rise of mass movements agitating for political participation. Achieving higher wages requires some fundamental changes in perspective from unbridled capitalism or authoritarian socialism to a variety of competitive, social democratic politics. It also calls for a civil society separate from government and the market forces to press for reform. In advanced capitalist systems, social legislation such as anti-trust, collective bargaining, unemployment insurance, minimum wage, and maximum labour hours laws, contributed to these building blocks of social democracy. In his 'theory of countervailing power', Galbraith (1993) recognized this phenomenon by arguing for Big Business to be countered by Big Labour, Big Consumer Unions, and Big Government. However, automation and robotization have already undermined the labour unions, while anti-government sentiments are attempting to dismantle the apparatus of the welfare state. Democratic regimes in the advanced industrial societies are therefore facing a serious crisis. The politics of anti-politics and a nostalgic, neoconservatism is currently trying to recapture the original impulses of competitive capitalism in an age of conglomerate capitalism, flexible accumulation, techno-logical robotization, structural unemployment, and social decay.

But in the case of the late-comers to industrialization such as Japan, the Soviet Union, China, or the two Koreas, the development of a civil society *vis-à-vis* the government and business sectors has lagged far behind. The state took up the challenge of 'catching up' at times in collaboration with business (as in Japan and South Korea) and at other times without (as in the Soviet Union and Red China before 1978). However, in order to catch up, levels of national savings have had to reach as high as 30 per cent in some of NICs as compared to the 5–10 per cent of GDP in the more leisurely pace of economic growth of Western Europe and North America. Modernization, however, inevitably leads to demands for political participation by those sectors of the population which have been denied increases in their wages and status. At this stage, the state can choose to democratize and raise wages or repress. Following long periods of keeping wages down by repressive regimes, the Soviet Union, China, Singapore, Taiwan, and South Korea are currently undergoing such processes of democratization. Conversely, in the face of a mobilized population, democratization can take place without modernization. In their post-independence phase of development, many of the LDCs faced such a dilemma. When resources are meager, the state can be both regressive and repressive, resulting in failures of both modernization and democratization. A more balanced approach to modernization can accommodate the progressive needs for democratization with the requirements of modernization by lowering expectations, leveling incomes, while increasing national savings and investment.

The role of communication in this process is one of integrating competing demands in a process of negotiating national goals and values. To the extent that they exist, freedoms of speech, assembly, and association provide a public sphere of discourse in which such conflicting demands for resources are negotiated and mediated democratically. The LDCs, however, face a serious dilemma by living in an age of global communication in which their people are exposed to the higher standards of life in the MDCs. Global communication and advertising expose them to rising political and economic expectations while introducing them to technological opportunities for production as well as consumption leap-frogging. Three basically different strategies present themselves. They may be labelled as isolationist, assimilationist, or participationist. The isolationist strategy attempts to close the society to the flows of international communication, as the Soviet Union and China did for a while, while paying for the costs of such isolation. The assimilationist strategy follows a more or less open-door policy, leading inevitably to assimilation in the vortex of global advertising. Iran under the Shah, Thailand, Algeria, and Egypt have followed such a permissive policy with dislocating consequences. In contrast, as Japan, South Korea, or Singapore have done, the participationist strategy would pick and choose its kinds and levels of international communication. The latter strategy has proved to be the most effective for scientific, technological, and economic leap-frogging while keeping the levels of desire, consumption, and democracy at bay.

Typically, however, the course of modernization is characterized by historical cycles from high accumulation to high mobilization while the processes of integration through the public sphere act as a mediating force. The development of democratic institutions reduces the severity of the cycles by allowing feedback mechanisms to correct the excesses of income inequalities that inevitably occur in the course of capital accumulation. In other words, communicative rationality and public discourse integrate society along more consensual patterns

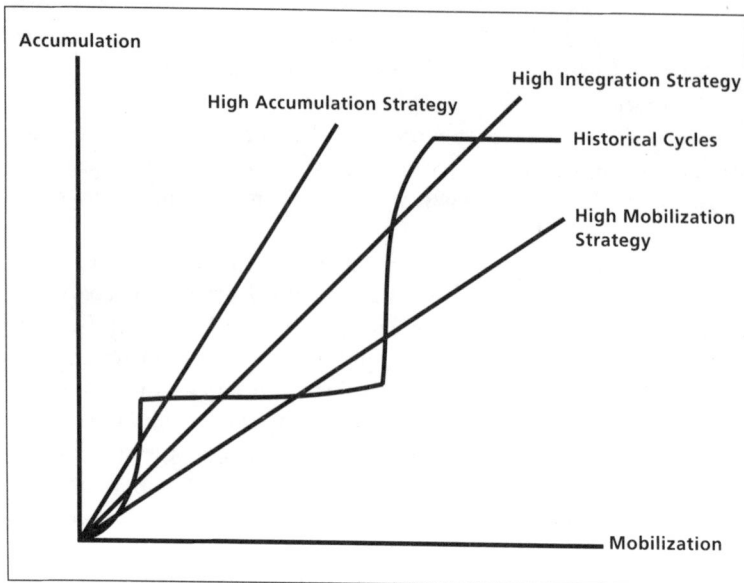

Figure 3: Historical cycles in modernization and democratization.
Source: Tehranian, 1990.

of progress. Figure 3 presents a schematic view of these historical cycles of high accumulation vs. high mobilization vs. high integration strategies. It demonstrates the basic policy options that face developed as well as developing countries. The trade-offs between capital accumulation and political mobilization are equalized only at the 45° line where a society can maintain its equilibrium by levels of cultural integration commensurate with its level of development (a high integration strategy). If, however, capital accumulation takes place at the expense of political participation (a high accumulation strategy), the forces of political mobilization will sooner or later assert themselves to demand greater political participation and more equal income distribution.

The modern industrial system has evolved into a succession of different types of capitalism, including the variety known as 'socialism' or 'state capitalism'. Authoritarian states have been perhaps an appropriate social and economic formation in the early stages of primitive accumulation when lack of infrastructural facilities, lumpy requirements of capital, and imperfections of the market often call for forced savings, and state planning and investment. However, as economic development reaches higher levels of accumulation and consumption, the complexity of the decisions needed to be made calls for a market orientation. The interplay of the forces of supply and demand in the market is far more effective in making the increasingly complex investment and consumption decisions. Moreover, the post-war rise of transnational corporations has facilitated the transfer of science, technology, capital, and management techniques to the less developed areas of the world. No Planning Commission or Gosplan can take the place of the numerous investors, producers, and consumers that run a modern, complex, industrial economy. As demonstrated by the rise of market economies in the formerly socialist countries, there is reason to believe that capitalism will continue its evolution and adaptation to new economic, socio-cultural, and environmental circumstances.

As the dramatic rise in suicides in the Soviet Union shows (Gerasimov, 1994), the transition from a command to a market economy is painful unless the social safety nets that liberal capitalism has developed are also built into the social structure. The demise of the Soviet Union and the growth of market economies in Eastern Europe, China, Vietnam, and Cuba may be viewed as continuing efforts in modernization, but democratization is not a necessary outcome. Witness the continuing repression in China and the recent electoral victories of the former Communists in many of the Eastern European and former Soviet republics. Similarly, the rise of religious ideologies has demonstrated that the processes of modernization are replete with twists and turns, including the possibility of theocracies that may become themselves carriers of modernization (Marty and Appleby, 1991, 1992, 1993; Juergensmeyer, 1993; Tehranian, 1993a & b). Contrary to the views of 'fundamentalism' that see it as a reaction against modernity, the current religious resurgence in the LDCs may be alternatively considered as disguised forms of modernization movements to mobilize deeply dislocated, traditional societies. After all, that was what the Reformation and its Protestant off-shoots accomplished for the Christian West. What form modernization will take in the future perhaps depends less on the inherent, cosmological features of conflicting civilizations and more on the economic, political, and cultural relations among the world centres and peripheries of power.

Robert J. Barro (1994) has argued that, in the long run, economic growth often brings forth democracy. From an empirical study of 100 countries in various stages of economic development from 1960 to

1990, his data reveal a linkage between economic development and the propensity to experience democracy. 'Non-democratic countries that have achieved high standards of living – measured by real per-capita GDP, life expectancy and schooling – tend to become more democratic over time. Examples include Chile, South Korea, Taiwan, Spain and Portugal. Conversely, democratic countries with low standards of living tend to lose political rights over time. Examples include most of the newly independent African states in the 1970s.' In this study, democracy is measured by the index of political rights compiled in the serial publication, *Freedom in the World*, edited until recently by Rayomd Gastil. Barro also forecasts that if their current rates of economic growth are maintained, some countries can be expected to achieve higher levels of democracy by the year 2000 (see Table 2).

Unfortunately, we are all dead in the long run. Authoritarian regimes often make fatal mistakes in the processes of transition to an industrial society. During 1958–61, nearly 30 million people died of famine in China under Mao because the leadership of the Communist

Table 2: If prosperity brings democracy

These countries will be more democratic...			*These countries will be less democratic...*		
	1993	*2000*		*1993*	*2000*
Iraq	.00	.21	Hungary	1.00	.81
Haiti	.00	.24	Mauritius	1.00	.81
Sudan	.00	.24	Botswana	.83	.66
Syria	.00	.32	Papua New Guinea	.83	.65
Algeria	.00	.33	Nepal	.83	.60
Swaziland	.17	.35	Bolivia	.83	.58
Iran	.17	.41	Bangladesh	.83	.56
Yugoslavia	.17	.41	Gambia	.83	.54
Indonesia	.00	.43	Benin	.83	.50
South Africa	.33	.47	Pakistan	.67	.48
Peru	.33	.51	Mali	.83	.44
Singapore	.33	.61	Congo	.67	.42
Taiwan	.50	.64	Niger	.67	.37
Hong Kong	.33	.67	Central African Republic	.67	.36
Mexico	.50	.72	Zambia	.67	.35

Note: The democracy index uses a scale from 0 to 1, where 0 means no political rights and 1 means virtually full rights. The figures for 1993 and other years are derived from information presented in Raymond Gastil and followers, *Freedom in the World*, various issues. Their subjective classifications follow the basic definition: 'Political rights are rights to participate meaningfully in the political process. In a democracy this means the right of all adults to vote and compete for public office, and for elected representatives to have a decisive vote on public policies'. The value shown for 2000 are projections based on the author's statistical analysis.
Source: Barro, 1994.

Party was unaware of what was going on in the countryside. In contrast, famines have been largely averted in the LDCs with some democratic channels of communication to government (J. Tehranian, 1995). For every enlightened dictatorship, we can name many more unenlightened ones characterized by incompetence, corruption, cult of personality, and edifice complexes. There are also many countries that enjoy relative prosperity but few democratic rights. South Africa before the dismantling of apartheid, South China, Indonesia, and Singapore come immediately to mind.

Ideologies of primitive accumulation tend to be ideologically primitive. They often hark back to the most primitive resentments in ethnic, religious, racial, or class discrimination. They have captured the imagination of millions in the rise of communism, fascism, and religious fundamentalism in the early or middle phases of industrialization. In the later phases of modernization, however, the ideologies of class privilege or resentment become far more complex. Witness the rise of a variety of élitist doctrines disguised in the form of 'religious' or 'scientific' legitimations of discrimination. In the United States, for instance, the religious right often focuses on the struggle against abortion, homosexuality, and separations of church and state while the scientific doctrines marshal evidence on behalf of natural inequalities in intelligence quotas (IQs). An example of the latter is the work of Herrnstein and Murray (1994) who have argued against public spending on health, education, and welfare on the grounds that such spending is wasted on those deficient in intelligence as measured by IQs (DeParle, 1994). The scientific findings about intelligence are extremely limited and tentative (Avery *et al.* 1994). The influence of genetic factors in determining intelligence is estimated to be anywhere between 40 to 80 per cent. Although there are significant differences among the different ethnic groups in the United States, no conclusion can be drawn on whether genetic or environmental factors or a mix of the two are responsible for the differences. What goes unnoticed in such studies is the fact that 'intelligence' can be manifested in a variety of forms and dexterities, including visual, mechanical, and perceptual abilities as well as the traditional three Rs of reading, writing, and arithmetic. As Thomas Edison declared, 'genius is 1 per cent inspiration and 99 per cent perspiration.' Motivation for hard work is probably a better predictor of success than IQs.

The prevailing ideological controversies on democratization are thus complex and ever-changing. Globalization of the world economy is creating a global cognitive élite that is directly or indirectly tied to the institutions of technocratic capitalism, driven by de-territorialization and informatization. The globalist ideology of liberal capitalism is, however, torn between the contesting principles of one-dollar-one-vote vs. one-person-one-vote. The semi-peripheries (the emerging middle classes) are clamouring for political democracy (in China, South Korea, Taiwan, Latin America), while the peripheries are resorting to the primordial, tribal identities of language (India), religion (fundamentalism), race (Africa, USA), ethnicity (old and new imaginaries), and gender (women) to press for equal rights. The neo-tribalist politics can be intolerant and totalitarian in orientation, but they have to be understood in the context of a world in which the economic and cultural survival of the LDCs are at risk.

Unless new strategies of modernization are adopted to reduce the gaps and provide bridges of trade, co-operation, dialogue, and understanding, the global encounter of materially uneven worlds of development can lead to clashes of cultures and civilizations. Civilizations are epistemological, knowledge systems. When two civilizations interact, they often develop a third civilization, culture, epistemology, and knowledge system. Out of the worldwide confluence of civilizations, we are developing a global civilization side by side with the old regional civilizations as well as national and local cultures. Dialogue is therefore the key to a successful development of a universal, human civilization in whose idiom we all need to speak in order to understand the national and local in their rich variety of human sub-cultures.

Democracy is a complex idea that should be considered as an agenda for a global negotiation of meaning. Competing inter-pretations have emphasized security, freedom, justice, or community. Clearly, democracy has to be contexualized and actualized within specific national and cultural traditions. Without global communica-tion on the meanings of democracy in different socio-historical contexts, we risk a cognitive tyranny by one set of countries against others. However, dialogical communication cannot be achieved without the conditions of freedom and equality in discourse. Unless and until such conditions obtain, international communication on democracy can be expected to be a dialogue of the deaf (Tehranian, 1982; Traber and Nordenstreng, 1992). And since violence often dictates under such conditions, it is fitting to remember Gandhi's view of democracy: 'My notion of democracy is that under it the weakest should have the same opportunity as the strongest. That can never happen except under non-violence' (Gandhi, 1948: 269).

References

Avery, Richard D. *et al.* (1994). 'Mainstream Science on Intelligence', *Wall Street Journal*, 13 December 1994, p. A18.

Anderson, Benedict (1983). *Imagined Communities: Reflections on the Origin and Spread of Nationalism*. London and New York: Verson.

Barnet, Richard J. and Roland E. Muller (1974). *Global Reach*. New York: Simon & Schuster.

Barro, Robert J. 'Democracy: A Recipe for Growth?', in *Wall Street Journal*, 1 December 1994, p. A14.

DeParle, Jason (1994). 'Daring Research or Social Science Pornography?', *New York Times Magazine*, 9 October 1994, pp. 48ff.

Eisenstein, Elizabeth L.(1979). *The Printing Press as an Agent of Change*, Vols. 1–2. Cambridge: Cambridge University Press.

Frankfort, Henri, *et al.* (1963). *Before Philosophy: The Intellectual Adventures of Ancient Man*. Baltimore, MD: Penguin Books.

Galbraith, J. K. (1993). *American Capitalism: The Concept of Countervailing Power*. New Brunswick, NJ: Transaction Publishers.

Galtung, Johan and Richard Vincent (1992). *Global Glasnost: Toward a New Information and Communication Order*. Cresskill, NJ: Hampton Press.

Gandhi, Mohandas K. (1948). *Non-Violence in Peace and War*, Vol. 1.

Gerasimov, Gennadi I. (1994). 'A Death Wish Is Haunting Russia', *New York Times*, 2 December 1994, p. 23.

Heilbroner, Robert L. (1962). *The Making of Economic Society*. Englewood Cliffs, NJ: Prentice Hall.

Herrnstein, Richard J. and Charles Murray (1994). *The Bell Curve*. New York: The Free Press.

Hobson, J. A. (1902). *Imperialism: A Study*. London: J. Nisbet.

Juergensmeyer, Mark (1993). *The New Cold War? Religious Nationalism Confronts the Secular State*. Berkeley: University of California Press.

Kaku, Michio (1994). *Hyperspace: A Scientific Odyssey Through Parallel Universes, Time Warps, and the Tenth Dimension*. New York: Oxford University Press.

Lenin V. I. (1917). *Imperialism: The Last Stage of Capitalism*. Moscow.

Lipset, Seymour M. (1963). *The First New Nation: The U.S. in Historical and Comparative Perspective*. New York: Basic Books.

Marty, Martin and Scott Appleby (eds.) (1991). *Fundamentalism Observed*. Chicago: Chicago University Press.

Marty, Martin and Scott Appleby (eds.) (1992). *Fundamentalism and State*. Chicago: Chicago University Press.

Marty, Martin and Scott Appleby (eds.) (1993) *Fundamentalism and Society*. Chicago: Chicago University Press.

Nye, Joseph (1990). *Bound to Lead: The Changing Nature of American Power.* New York: Basic Books.

Riggs, Frederick (1994). 'Ethnonationalism, industrialism, and the modern state', *Third World Quarterly*, 15:4, 583–611.

Robertson, Roland (1994). *Globalization: Social Theory and Global Culture.* London: Sage.

Robertson, Roland (1994). 'Globalisation or Glocalisation', *Journal of International Communication*, 1:1, June 1994, 33–52.

Rowe, C. and F. Koetter (1992). *Collage City.* Cambridge, MA, MIT Press.

Tawney, R. H. (1984). *Religion and the Rise of Capitalism: A Historical Study.* New York: Penguin Books.

Tehranian, John (1995). 'The Political Economy of African Famine: Ethiopia, Kenya, and Zimbabwe, 1982–5', unpublished BA thesis, Harvard University.

Tehranian, Katharine (1994). *Modernity, Space, and Power: The American City in Discourse and Practice.* Cresskill, NJ: Hampton Press.

Tehranian, Katharine and Majid Tehranian (eds.) (1992). *Restructuring for World Peace: On the Threshold of the 21st Century.* Cresskill, NJ: Hampton Press.

Tehranian, Majid (1982). 'International Communication: A Dialogue of the Deaf?', *Political Communication and Persuasion*, 2:2.

Tehranian, Majid (1990). *Technologies of Power: Information Machines and Democratic Prospects.* Norwood, NJ: Ablex Publishing Corporation.

Tehranian, Majid (1992). 'Khomeini's Doctrine of Legitimacy', in Anthony J. Parel and Ronald C. Keith (eds.), *Comparative Political Philosophy.* New Delhi: Sage Publications.

Tehranian, Majid (1993a). 'Ethnic Discourse and the New World Dysorder', in Colleen Roach (ed.), *Communication and Culture in War and Peace.* Newbury Park: Sage Publications.

Tehranian, Majid (1993b). 'Fundamentalist Impact on Education and the Media: An Overview', in Martin E. Marty and R. Scott Appleby (eds.), *Fundamentalism and Society.* Chicago: University of Chicago Press.

Traber, Michael and Kaarle Nordenstreng (1992). *Few Voices, Many Worlds: Towards a Media Reform Movement.* London: World Association for Christian Communication.

Trachtenberg, Alan (1982). *The Incorporation of America: Culture and Society in the Guilded Age.* New York: Hill & Wang.

Tocqueville, Alexis de (1956). *Democracy in America*, abridged and edited by R. D. Heffner. New York: Mentor Books.

Toffler, Alvin (1970). *Future Shock.* New York: Bantam Books.

Weber, Max (1958). *The Protestant Ethnic and the Spirit of Capitalism*, trans. by Talcott Parsons. New York: Scribner.

Woodall, Pam (1994). 'A Survey of the Global Economy', *Economist*, 1 October 1994, pp.1–38.

4

Communication ethics as the basis of genuine democracy

CLIFFORD G. CHRISTIANS

Let me summarize my argument in a paragraph. The Enlightenment was besieged by a dichotomy that still needs resolution today for genuine democracy to prosper. Important intellectual debates about the origins of modernism centre that split on either the subject–object, material–spiritual, or fact–value dichotomies. Contrary to those interpretations, I believe the Enlightenment mind could not integrate freedom with the moral order, and this perennial human dilemma remains to be solved. Communication studies can contribute to this integration by articulating a holistic view of truth in moral rather than epistemological terms. In this respect, Michael Traber's work is of historic value, because it overcomes the modernist dichotomy between freedom and a moral universum.

Enlightenment dichotomies

The Enlightenment is the decisive modern revolution.[1] Nothing has been so formative of the Western mind. And to paraphrase Oscar Wilde, whoever does not know it well is condemned to repeat it. The intellectual revolutions of the previous two centuries – that is, the Age of Reason and the Age of Science – exploded into this audacious and entangled historical watershed. Most of the writers and thinkers were popularizers, journalistic types centred largely in France and known as *philosophes*: François Marie Arouet (Voltaire), Jean Jacques Rousseau, Denis Diderot, Baron Holbach, and Antoine Caritat Marquis de Condorcet.

Book publishing on the European continent underwent an astonishing transformation. As the eighteenth century dawned, nine out of ten books appeared in Latin and were available only to the intelligentsia; a century later, eight out of ten were printed in the vernacular instead. Knowledge was disseminated on an unprecedented scale; literacy rates doubled and a learned class was born. It was the century of the German geniuses in music – Johann Sebastian Bach, George Frederic Handel, Franz Joseph Haydn, and Wolfgang Amadeus Mozart. The American statesmen Benjamin Franklin and Thomas Jefferson were quintessential figures in the movement. Edward Gibbon penned his vitriolic *Decline and Fall of the Roman Empire* during this period. And befitting an age which extolled human centrality and historical progress, portraiture was adopted by the wealthy class as the preferred form of art. Given its magnitude, the collapse in our own day of the Enlightenment worldview certainly generates earthquake shocks everywhere.

The Enlightenment mind clustered around an extraordinary dichotomy. Intellectual historians usually summarize this split in terms of subject–object, fact–value, or material–spiritual dualisms. And all three are legitimate interpretations of the cosmology inherited from Galileo, Descartes, and Newton. However, communication scholars addressing our crisis age must enter this scholarly debate with a revisionist purpose, recognizing the importance of these typical dualisms but identifying a fourth as more earth-shaking than the others for the prospects of genuine democracy.

The Enlightenment story actually begins in the sixteenth century with the Italian Galileo Galilei (1564–1642), a central figure in the transition from medieval to modern science. Galileo mapped reality in a new way, dividing nature into two famous compartments – primary: matter, motion, mass, mathematics; and secondary: the *meta*physical, *super*natural, values, meaning. In *The Assayer,* Galileo writes, 'This great book, the Universe . . . is written in the language of mathematics, and its characters are triangles, circles, and geometric figures' (Galilei, 1957 ed.: 238–9). Matter alone mattered to him; he considered all non-material immaterial. He separated off the qualitative as incapable of quantitative certainty. In effect, he suggested two essences – values and meaning on the one hand, matter and quantity on the other. His fascination with the Copernican world-picture motivated him to promote heliocentricity not as the calculation of astronomers only, but as a wide-ranging truth about the structure of reality. Lewis Mumford (1970) in his *Pentagon of Power* ridicules 'the crime of Galileo', because his bifurcation allowed the world of value and meaning to start shrivelling away.

Within a century, the Englishman Isaac Newton could describe the world in his *Principia Mathematica* (1687) as a lifeless machine composed of mathematical laws and built on uniform natural causes in a closed system. The upper story had been dissolved. Phenomena could be explained as the outcome of an empirical order extending to every detail. Mystery was defined away. All but quantity or number were called sophistry and illusion. Principles of mass and gravity extended to the extreme limits of the cosmos – explaining the movement of the farthest planets with the same mathematical laws as described an apple dropping from a tree. Newton provided mature formulations, raising the mechanistic worldview to an axiomatic, independent existence.

Ironically Newton was committed to the upper story that his *Principia* eroded. During his lifetime, he wrote 1.3 million words on theology, mastered the writings of the early church fathers, and was a generous supporter of Anglican church projects around London. Yet among his scientific colleagues, he banned any subject touching the sacred, insisting that 'we are not to introduce divine revelations into science, nor philosophical opinions into religion.' Newton's loyalties were firmly anchored in both scientific method and transcendent truth. His Enlightenment heirs, however, would abandon the upper story with the same zeal that Newton applied to founding new churches. Voltaire (1694–1778), for example, pushed the material–spiritual split to its extreme, at least in his prolific contributions to the *French Encyclopedia* if less so in *Candide*.

Within that pattern from Galileo to the pervasive scheme of Newton stands René Descartes (1596–1650), who cut the dichotomy firmly into the being of *homo sapiens*. Although Galileo and Newton inspired the Enlightenment as much as anyone, the Frenchman René Descartes contradicted most vehemently a holistic view of reality and ensured that persons also would be swept into the new cosmology. Descartes insisted on the non-contingency of starting points. He presumed clear and distinct ideas, objective and neutral, apart from anything subjective.

Consider the very conditions under which Descartes wrote *Meditations II* in 1642. The Thirty Years War was spreading social chaos throughout Europe. The Spanish were ravaging the French provinces and even threatening Paris. But Descartes was in a room in Belgium on a respite, isolated in seclusion. For two years even his friends could not find him hidden away studying mathematics. Tranquillity for philosophical speculation mattered so much to him, that upon hearing that Galileo had been condemned by the Roman Catholic Church he retracted parallel arguments of his own on natural science.

His *Discourse on Method* (1637) elaborates this objectivism in more detail. Genuine knowledge is built linearly, with pure mathematics the least touched by circumstances. Two plus two equals four was lucid and testable, and all genuine knowledge in Descartes' view should be as cognitively clean as arithmetic. In *Rules for the Direction of the Mind*, Descartes contended, in effect, that one could demonstrate truth only by measurement. Therefore, he limited his interest to precise, mechanistic, mathematical knowledge of physical reality. As E. F. Schumacher has complained, no one sketched the modern intellectual map more decisively than Descartes, and his philosophical map-making defined out of existence those vast regions which had engaged the intense efforts of earlier cultures and non-Western peoples.

Neither the subject–object or material–spiritual or fact–value split puts the Enlightenment into its sharpest focus, however. Its deepest root was a pervasive autonomy. What prevailed was the cult of human personality in all its freedom. Human beings were declared a law unto themselves, set loose from every faith that claimed their allegiance. Proudly self-conscious of human autonomy, the eighteenth century mind saw nature as an expansive arena for exercising freedom, a field of limitless possibilities in which the sovereignty of human personality was demonstrated by its mastery of the natural order. Release from nature spawned autonomous individuals who considered themselves independent of any authority. The freedom motif – persons understood as ends in themselves – was the deepest driving force, first released by the Renaissance and achieving maturity during the Enlightenment.

Jean-Jacques Rousseau was the most outspoken advocate of this radical freedom. He gave intellectual substance to free self-determination of the human personality as the highest good. Rousseau contended that freedom embodied in human beings justified itself as the final aim. 'I long for a time', he wrote, 'when freed from the fetters of the body, I shall be myself, at one with myself . . . when I myself shall suffice for my own happiness' (Rousseau, 1961: 350). Although such unbridled liberty had taken root earlier with Pico de Mirandolla's 'Oration on the Dignity of Man' in the Renaissance, the Swiss Rousseau provided its most mature version. In

Emile (1762), for instance, he contended that civilization's artificial controls demean humanity and make us vicious, whereas free in a state of nature the 'noble savage' lives in harmony and peace. As Joseph de Maistre observed, for Rousseau, asking why people born free were nevertheless everywhere in chains was like asking why sheep born carnivorous everywhere nibbled grass. What we observe empirically does not invalidate our own true nature. Liberty is the inalienable ingredient that makes humans human, even though under so-called normal conditions its sacred frontiers are desecrated.

Rousseau is a complicated figure. He refused to be co-opted by Descartes' rationalism, Newton's mechanistic cosmology, or Locke's egoistic selves. And he was not merely content to isolate and sacralize freedom either, at least not in his *Discourse on Inequality* or in the *Social Contract* where he answers Hobbes. His conclusion that a collective can be free if it enacts its own rules which are then obeyed voluntarily, is only a partial solution – though superior to more static contractarian theories and a champion of popular sovereignty. In distinguishing the general will from the empirical will of all, Rousseau, of all the Enlightenment heavyweights, recognized that freedom and the moral order feed off one another, at least in principle.

Rousseau represented the romantic wing of the Enlightenment, revolting against its rationalism. He won a wide following well into the nineteenth century for advocating immanent and emergent values rather than transcendent and given ones. While admitting humans were finite and limited, he nonetheless promoted a freedom of breath-taking scope – not just disengagement from God or the Church, but freedom from culture and from any authority. Even among those with a less pastoral vision, autonomy became the core of human being and the centre of the universe. Rousseau recognized the consequences more astutely than those comfortable with a shrunken negative freedom. But the only solution that he found tolerable was a noble human nature which enjoyed freedom beneficently and, therefore, one could presume, lived compatibly in some vague sense with a moral order. His understanding of equality, social systems, axiology, and language were not finally anchored in an adequate philosophical anthropology.

Obviously one can reach autonomy by starting with the subject–object dualism. In constructing the Enlightenment worldview, the prestige of natural science – then typically called 'natural philosophy'

– played a key role in setting people free. Achievements in mathematics, physics, and astronomy allowed humans to dominate nature which formerly had dominated them. In Cartesian terms, the scientific method enabled the human race to be 'masters and possessors of nature'. Science provided unmistakable evidence that by applying reason to nature and human beings in fairly obvious ways, people could live progressively happier lives. Crime and insanity, for example, no longer needed repressive theological explanations, but were deemed capable of mundane empirical solutions. By characterizing the problem as primarily epistemological, one tends to find the post-Enlightenment alternative in epistemology. The burning issue then becomes how we can know – and all kinds of subjectivity models or phenomenology or contemporary hermeneutics are directed precisely towards overcoming the scientistic notion of lawlike abstractions and operational definitions through fixed procedures.

Likewise one can get to the autonomous self by casting the question in terms of a radical discontinuity between hard facts and subjective values. The Enlightenment did push values to the fringe by its disjunction between knowledge of what is and what ought to be. And Enlightenment materialism in all its forms isolated reason from faith, knowledge from belief. As Robert Hooke insisted three centuries ago when he helped found London's Royal Society: 'This Society will eschew any discussion of religion, rhetoric, morals, and politics.' With factuality gaining a stranglehold on the Enlightenment mind, those regions of human interest which implied oughts, constraints and imperatives simply ceased to appear. Certainly those who see the Enlightenment as separating facts and values have identified a cardinal difficulty. Likewise, the realm of the spirit can easily dissolve into mystery and intuition. If the spiritual world contains no binding force, it is surrendered to speculation by the divines, many of whom accepted the Enlightenment belief that their pursuit was ephemeral.

But the Enlightenment's autonomy doctrine created the greatest mischief. Individual autonomy stands as the centrepiece, bequeathing to us the universal problem of integrating human freedom with moral order. This perennial question appears on the human agenda in various forms: determinism and free will, constraint and emancipation, order and anarchy, the liberty of conscience, dynamic socialization and stultifying institutions, ideology and praxis, freedom and responsibility.

But whatever its specific formulation, the nexus of human freedom and moral order remains a classic concern for the philosophical mind. And in struggling with the complexities and conundrums of this relationship, the Enlightenment, in effect, refused to sacrifice personal freedom. Even though the problem had a particular urgency in the eighteenth century, its response was not resolution but categorically insisting on autonomy. Given the despotic political regimes and oppressive ecclesiastical systems of the period, such an uncompromising stance for freedom at this juncture is understandable. The Enlightenment began and ended with the assumption that human liberty ought to be cut away from the moral order, never integrated meaningfully with it. To be successfully counter-Enlightenment, we must take a radical stance precisely at this point; if this dichotomy remains unresolved a democratic life of *eudaemonia* and *shalom* is totally impossible.

Alexis de Tocqueville's monumental *Democracy in America* recognized the consequences of failing to integrate individual freedom with moral norms. While democracy, in his view, operates with an egalitarian language in everyday affairs, it needs a deeper level – that is, forms of moral discourse which anchor the self and society, and provide a sense of calling for our jobs, leisure, and politics. These social integuments Tocqueville saw as moderating democracy's more destructive potentialities. As a Catholic, he believed that liberty needed religion to mitigate those excesses that threatened its survival. He argued for making moral sense of our lives rather than pursuing the unencumbered self. In this appeal, Tocqueville read the Enlightenment problematic correctly and reaffirmed that unordered liberty becomes licence. He would not be surprised one century later that we are overcome today by the banality of a social order exacted in the name of fulfilment. Life together has become not a struggle for social change but for self-realization. Our current unseemly self-gratification, our narcissism, has rendered public life virtually impossible and simultaneously hollows out our personal sphere as well.

When the Enlightenment gave birth to a reductionist but distinctive freedom, pretending the self could be separated from nature and from culture, the very possibility of understanding basic human questions was erased in principle. Modernity is unable to negotiate moral criteria or understand the nature of community and moral judgement because autonomy pervades its ethos. As Daniel Callahan concludes:

Autonomy should be a moral good, not a moral obsession. It is *a* value, not *the* value. If . . . it rests on the conviction that there can be no common understanding of morality, only private likely stories, then it has lost the saving tension it competitively needs with other moral goods . . . I am told that I have the right to fashion my own moral life and shape my own moral goals. But how do I go about doing that? . . . Autonomy, I have discovered, is an inarticulate bore, good as a bodyguard against moral bullies, but useless and vapid as a friendly, wise, and insightful companion.

(Callahan, 1984: 42)

Isolating freedom eclipses the integral substance of morality at the starting point. Our ethical discourse becomes gravely disordered – the emotivist self, autonomism in morality, and fictions like natural rights. Genuine democratization of empowered citizens cannot occur unless freedom and the moral order are reintegrated.

Most philosophers of history recognize that the Enlightenment age has now run its course. Eastern Europe and the USSR are forever changed. The West is finishing a historic period also; we are only left to debate the appropriate nomenclature – postmodern, post-liberal, post-factual, post-colonial, post-structural, or post-patriarchal. Today's gratuitous hedonism, technocratic rationality, and debilitating secularism are the Enlightenment at its ragged edge. Its conceptual inconsistencies have finally been exposed and largely discredited. Underneath the shrill rhetoric and often overwrought claims, the West recognizes this age as a *chairos*, a strategic moment, a defining time in world affairs.

The cancerous effect of the freedom/moral order dualism has been long-term and finely grained, but now inescapable. On the one side, the divines – themselves relegated to the fringes – maintain guardianship over moral scraps. On the other, the few serious struggles have been transmogrified into epistemological discourse. Early on the ethical question of how we should live floundered on debates over whether it was cognitively meaningful, which itself succumbed to philosophical relativity, until today universal norms are largely alien to the Western mind.

Truth as master norm

However, as the curtain comes down on the Enlightenment era and another episode takes shape on the stage of history, the debate over freedom and the moral order will not disappear. Were it not for Enlightenment hubris, it would have been recognized that this is a permanent issue which never fades from the human agenda and can never be totally resolved on any occasion. Certainly the connection between freedom and responsibility demands urgent attention by those of us committed to forms of communication that enable genuine democracy to prosper in the civic order. The history of modernity reminds us that eliminating the subject/object dualism, piously asserting value-centredness, and rejecting excessive materialism are all achievements in themselves, but of little consequence if freedom remains isolated from any overarching standards. The press's objectivism, instrumentalism, and technicism are all rooted in an Enlightenment paradigm gone to seed. Exorcizing these demons is only a Pyrrhic victory if liberty and morality are not conceived as one organic whole. Our efforts at genuine community will be futile as long as democracy's infrastructure is hoist by the Enlightenment's petard.

Richard Rorty scourges the Enlightenment worldview as 'the quest for certainty over the quest for wisdom'. Modernist philosophers, have sought:

> . . . to attain the rigor of the mathematician or the physicist, or to explain the appearance of rigor in these fields, rather than to help people attain peace of mind. Science, rather than living, became philosophy's subject, and epistemology its centre. (Rorty, 1979: 365–94)

Rorty labels this epistemological system 'foundationalism', though 'objectivism ' is a more typical name for the amalgam of practices and commitments that have prevailed in Western thought as a whole.

Talcott Parsons has been America's most influential sociologist of this century. He insisted on the objectivist credo while helping smuggle in Nazi collaborators as Soviet experts after the Second World War.[2] In the classroom, as head of the American Sociological Association, while training the next generation's leaders, this highly abstract functionalist pursued scientific status through value-free neutrality. Meanwhile, Parsons co-operated with the Russian military to bring Vladimir Pozdniakov, Leo Dudin of the University of Kiev, and Nicholas Poppe of Leningrad (war criminals all) to Harvard's Russian Research Center. It was self-evident to him that America's Cold War frenzy and Harvard's own George F. Kenan in the state department would guarantee unlimited funding for the Center if its expertise were as close to the ground as these Russian exiles represented. Apparently their knowledge of ethnic groups in Soviet Asia, once vital to the Nazi pursuit of Jewish communities, would now serve American ideology. It was the summer of 1948 as Parsons shuttled across the North Atlantic. Structural-functionalism purged of moral quandaries in Massachusetts engaged in reprehensible social practice in Germany. Scepticism and detachment – aloof from discourse that shapes the moral landscape – disinterested pursuit of truth, in Parsons' hands destroyed the very liberty it cherished for itself.

The attacks on this misguided view of human knowledge had already originated in Giambattista Vico's *fantasia* and Wilhelm Dilthey's *verstehen* in the Counter Enlightenment; they continued with American pragmatism, critical theory in the Frankfurt School, hermeneutics, and Wittgenstein's linguistic philosophy; until our own day when the phenomenal interest in Rorty, Thomas Kuhn, Michael Polanyi, interpretive research, and deconstructionism symbolizes a crisis in correspondence views of truth. Institutional structures remain Enlightenment driven, but in principle the tide has turned currently toward restricting objectivism to the territory of mathematics, physics, and the natural sciences. In reporting, objectivity has become increasingly controversial as the working press' professional standard, but it will remain entrenched in our ordinary practices of news production and dissemination until an alternative mission for the press is convincingly formulated.

The press under Enlightenment tutelage maintains representational accuracy as its *telos,* with adjustments in detail but not in principle. Though without the enthusiasm of earlier decades, we still presume that news corresponds to reality and is ideally bound to neutral algorithms. We counsel each other to make the best possible attempt at value-free reporting, even though never perfectly attainable. By analogy, we are told, a surgeon who cannot ensure an operating room free from bacteria, does not use a kitchen table and a butcher's knife. Objectivity is still the centrepiece of most journalism codes of ethics, and a majority of reporters continue to equate ethics with impartiality.

But rather than maintain a façade of objectivism – reporters as impersonal transmitters of facts – we need to articulate a fulsome concept of truth as communication's master principle. As the norm of healing is to medicine, critical thinking to education, craftsmanship to engineering, justice to politics, and stewardship to business, so truthtelling in its fulsome sense becomes the news profession's occupational norm. I intend this as a normative framework of a radically different sort, one that fundamentally reorders the news media's professional culture and enables it to serve democratization. Instead of an information enterprise trapped in the epistemological domain, truth should be relocated in the moral sphere.[3] Driving the modernist project is an objectivist way of knowing, centred on human rationality and armed with the scientific method. In the Enlightenment worldview, facts mirror reality. It aims for 'true, irrefragable [incontrovertible] accounts of an objective reality that is separate and different from human consciousness' (McKinzie, 1994: 33). In Bertrand Russell's formula, 'truth consists in some form of correspondence between belief and fact' (Russell, 1912: 121).[4] As Mark Johnson demonstrates, this illusion of context-free rationality poses great obstacles for morality; moral principles are allegedly derived from the essential structure of a disembodied reason. Rather than prizing care, reciprocity and imaginative ideals, our moral understanding becomes prescriptivist, rules oriented, and absolutist. 'What results is an extremely narrow definition of what counts as morality; . . . it is only doing the right thing; . . . it consists in discovering and applying moral laws. This drastic narrowing of the scope of morality has monumental consequences' (Johnson, 1993: 246). Truth, for example, is conceived in elementary epistemological terms as accurate representation; in this truncated form it makes no robust, contextual, social contribution to our public philosophy.

Rorty understands the significant stakes here, defining truth not as a 'mirror of nature' and 'privileged contact with reality,' but 'what is better for us to believe' (Rorty, 1979: 10). Since Walter Lippmann distinguished news and truth in the 1920s, the epistemology of news has been critiqued and debated, but truthtelling still has not received its due. Truth is a problem of axiology rather than epistemology. It belongs in the moral sphere and therefore should become the province of ethicists, especially when the dominant objectivist scheme has reached a historical crossroads.

When truth is articulated in terms of the moral order, we can mould its richly textured meaning around the Hebrew *emeth* (trustworthy, genuine, dependable, authentic), the Greek *aletheia* (openness, disclosure), the Serbo-Croatian justified (as plumbline true in carpentry). Dietrich Bonhoeffer's *Ethics* contends correctly that a truthful account lays hold of the context, motives, and presuppositions involved (Bonhoeffer, 1955: ch.5). Telling the truth depends on the quality of discernment so that penultimates do not gain ultimacy. Truth means, in other words, to strike gold, to get at 'the core, the essence, the nub, the heart of the matter'.[5] In Anthony Giddens' phrase, it entails 'discursive penetration' (Giddens, 1979: 73). For Henry David Thoreau – though addressing a different issue – when we are truthful, we attempt to 'drive life into a corner and . . . if it proves to be mean, why then to get the genuine meanness out of it and publish its meanness to the world; or if it were sublime, to know it by personal experience and be able to give a true account of the encounter' (1975: 94).

Augustine (AD 354–430), professor of rhetoric at Milan and later Bishop of Hippo, illustrates my intentions here. His rhetorical theory represents a major contribution to the philosophy of communication, contradicting the highly functionalized, secular and linear view bequeathed by the ancient Greeks. As with Aristotle, rhetoric entails reasoned judgement for Augustine; however, he 'break[s] away from Graeco-Roman rhetoric, moving instead toward . . . rhetoric as aletheiac act'.[6] Rhetoric for him is not knowledge-producing or opinion-producing but truth-producing (*aletheiac*).[7] In the *Epistolae* he challenges us to speak truthfully rather than cunningly. *De Doctrina Christiana* scourges the value-neutral, technical language of 'word merchants' without wisdom.[8] Truth is not fundamentally a prescriptive statement. The aletheiac act in Augustine 'tends to be more relational than propositional, a dialogically interpersonal, sacramentally charitable act rather than a statement . . . taking into account and being motivated by [the cardinal virtues] faith, hope, and charity' (Settle, 1994: 49, 57). The truth for him does not merely become clear, but motivates. In truthful communication for Augustine, 'it is not enough to seek to move men's minds, merely for the sake of power; instead, the power to move is to be used to lead men to truth' (Murphy, 1974: 62).

Augustine's searing critique of autonomous rationality was so penetrating that Arthur Kroker and David Cook credit him with setting the standard for cultural analysis until today. In contrast to postmodernism's rupture and against nothingness as the ultimate commitment, this 'Columbus of the modern experience', fashioned a normative domain by reconceiving truth as reason radiated by love (*caritas*) (Kroker and Cook, 1986: 37). 'Not only is *caritas* the goal of interpretation, it is also the only reliable means of interpretation'

(O'Donnell, 1985: 25). *Caritas* 'informs and directs the rhetorical process' (Settle, 1994: 56), or in St Paul's terms, 'love rejoices in the truth' (I Corinthians 13: 6). Conversely, solidarity with our neighbours is only possible over the long term when communicating virtuously, that is, when 'speaking the truth in love' (Ephesians 4: 15). Augustine subverts contemporary discourse while retaining a constructive ambience that links truth with moral principles.

Perhaps truth has languished so long in the epistemological desert that our social communication cannot be emancipated from facts and accuracy. Meanwhile, the power and domination literature insists that no appropriate conception of truth is possible anyhow under prevailing conditions of systematic distortion and repression.[9] For Jacques Derrida, modern discourse is an arbitrary system of differences, of oppositions and conventions; language is an unending series of significations allowing dogma and official codes to govern human existence. Linguistic games are said to supercharge the contemporary age, fragmenting it toward oblivion. And how realistic are we in demanding multi-layered explanations from a public medium such as television, constrained by its technology to visual immediacy?

For William James truth happens to an idea; and with Pilate scoffing at truth, Zen meditation seeking it in everyday life, Hegel believing in the truth of the organic *Gestalt,* Brunner insisting on truth as encounter, and Nietzsche calling it a social product, the serpentine entanglements of this pregnant term may no longer permit it to be a contemporary beacon. We have not even successfully identified as yet the distinctions that make a difference within truth's semantic field, though interpretative studies and hermeneutics (Gadamer's *Truth and Method* and Ricoeur's *History and Truth,* for instance) help provide orientation and specificity.

Furthermore, the idea of a moral order in which to situate truth is only in an embryonic stage. But one particular step forward is exceptional, *Meaning and Moral Order,* by Princeton sociologist Robert Wuthnow (1987). He argues that all cultures maintain a complex territory along the boundaries between the intentional and inevitable, actual behaviour and our aspirations, the conscience and socially constructed mores (Wuthnow, 1987: 71–5). This value-centred domain – dramatized by ritual and making human life purposive – is a 'relatively observable set of cultural codes' (95).[10] At epiphanal moments we enter this arena suspended outside our *persona,* symbols enabling us to reside simultaneously in the internal and external (Nagel, 1986). Thus Vaclev Havel and other politicians, parents with their children, educators working with students on a philosophy of life, social activists with integrity such as Michael Traber – all appeal connotatively to a moral order beyond themselves when they insist that the truth is non-negotiable. Language inflects truth statements *in locis*, yet at the same time symbolization situates truth claims outside our subjectivity.

With first foundations no longer a credible anchor for our ethical principles, Robert Wuthnow, Paul Ricoeur, Thomas Nagel and their circle are attempting to locate the latter as conditions of our humanness. In other words, if absolutes over time are inconceivable on this side of Newton, can universals be established across human space? The intellectual task of the post-Enlightenment is embedding normative principles, such as truth, in history, rather than presuming modernist metaphysics. Emmanuel Levinas is but one example of how to proceed. In his classics, *Totality and Infinity* (1969) and *Ethics and Infinity* (1985), rhetorical ethics breaks free from tyranny and violence by inscribing itself in the inexhaustible Other. Infinity exceeds its container in finite being. Infinity cannot be grasped by human reach or understood through human reason, though we desire it totally. We are transfigured through the Other as unfathomable difference; a third party arrives in our face-to-face encounters – the presence of the whole of humanity. In responding to the Other's need, a baseline for justice is established across the human race. Ethics is no longer a vassal of philosophical speculation, but is rooted in human existence. We seize our moral obligation and existential condition simultaneously.[11]

Epilogue
The nature of language was one fascination of eighteenth-century intellectuals. Some were preoccupied with the lingual *per se*, that is, with etymology, syntactics and phonetics. But more importantly for the issues in this volume, the Enlightenment as a whole understood the centrality of language in human affairs. For two centuries the West has benefited from and advanced the notion that language is the matrix of community, the catalytic agent in social formation. Thus the vision of a more democratic international order inevitably means revolutionizing our communications systems. When truth with moral significance becomes communication's defining feature, the global

community has at least the basic resources for peace, solidarity, mutual respect, and equality. In Jewish wisdom, truth is tied together with mercy (Genesis 32: 10), with mercy and justice (Isaiah 16: 5), and with peace (Zechariah 8: 16). In Psalm 85: 10, 'Mercy and truth will meet, justice and peace will kiss each other.' Within these linkages, truth is foundational: 'Justice is turned back, and righteousness stands afar off; for truth has fallen in the public squares, and righteousness cannot enter' (Isaiah 59: 14–15).[12] Michael Traber's books, editorials, and addresses bring this prophetic legacy into its own.

We face what Jürgen Habermas calls a crisis of legitimation. What counts as validity after post-structuralism? It is far from settled whether a credible version of normative values in general, and in truthtelling in particular, can be established without assuming an Enlightenment cosmology. But this I consider to be a worthwhile challenge for reflective ethicists, who believe that genuine democracy rests on moral principles. For students and practitioners of communication, recovering truth as a master norm is preferable to allowing the public media to lurch along through a post-factual modernity with an empty centre, while we put our scholarly energies into their short-term predicaments.

Notes

1. For an elaboration of the Enlightenment's impact on the mass media and public philosophy, see Christians, Ferré and Fackler, (1993), chs. 2, 4, 6. This section is summarized largely from pp. 18–25.
2. Cf. Jon Wiener (1989), 'Talcott Parson's Role: Bringing Nazi Sympathizers to the US,' pp. 1, 306–9.
3. Research on the moral concept of truth was made possible through the support of the Pew Evangelical Scholars Program.
4. For a summary of correspondence views, including both Russell's and J. L. Austin's versions, see Kirkham (1992), ch. 4.
5. Pippert (1989), *An Ethics of News: A Reporter's Search for Truth*, p. 11, for an initial attempt to define journalism in terms of truth.
6. Settle (1994), 'Faith, Hope, and Charity: Rhetoric as Aletheiac Act in *On Christian Doctrine*', p. 49.
7. This terminology is from Sullivan (1992), pp. 317–32.
8. Marjorie Boyle's 'Augustine in the Garden of Zeus,' (*Harvard Theological Review*, X, 1990, pp. 117–39), and the mini-classic by Kenneth Burke, *The Rhetoric of Religion: Studies in Logology* (Berkeley: University of California Press, 1970) provide a substantial bibliography on Augustine and rhetoric.
9. But refer to Anthony Giddens' effort (*The Constitution of Society:*

Outline of the Theory of Structuration, 1984) to reunite power and truth.
[10] See Paul Ricoeur (1973), pp. 153–65, for a similar argument.
[11] Levinas's work is elaborated in three papers presented at the Third National Conference on Ethics, Gull Lake, Michigan, May 1994: Wesley Avram, 'Discourse and Ethics at the "End" of Philosophy: The Case for Levinas'; Kenneth R. Chase, 'Rethinking Rhetoric in the Face of the Other'; and Jeffrey Ediger, 'The Ethics of Addressability in Emmanuel Levinas'.
[12] Summarized from Pippert (1989), p. 12.

References

Bonhoeffer, Dietrich (1955). *Ethics,* trans. N. H. Smith. New York: Macmillan.
Callahan, Daniel (1984). 'Autonomy: A moral good, not a moral obsession', *The Hastings Center Report*, October.
Christians, Clifford G., John P. Ferré and P. Mark Fackler (1993). *Good News: Social Ethics and the Press*. New York: Oxford University Press.
Galilei, Galileo. *Discoveries and Opinions of Galileo*, trans. Stillman Drake (1957). New York: Doubleday.
Giddens, Anthony (1979). *Central Problems in Social Theory*. Berkeley CA: University of California Press.
Giddens, Anthony (1984). *The Constitution of Society: Outline of the Theory of Structuration*. Cambridge: Polity Press.
Johnson, Mark (1993). *Moral Imagination: Implications of Cognitive Science for Ethics*. Chicago: University of Chicago Press.
Kirkham, Richard L. (1992). *Theories of Truth: A Critical Introduction*. Cambridge, MA: MIT Press.
Kroker, Arthur and David Cook (1986). *The Postmodern Scene: Excremental Culture and Hyper-Aesthetics*. New York: St Martin's Press.
Levinas, Emmanuel (1969). *Totality and Infinity: An Essay on Exteriority*, trans. Alphonso Lingis, Pittsburgh, PA: Duquesne University Press.
Levinas, Emmanuel (1985). *Ethics and Infinity: Conversations with Philippe Nemo*, trans. Richard A. Cohen. Pittsburgh PA: Duquesne University Press.
McKinzie, Bruce W. (1994). *Objectivity, Communication, and the Foundation of Understanding*. Lanham MD: University Press of America.
Murphy, James J. (1974). *Rhetoric in the Middle Ages: A History of Rhetorical Theory from Saint Augustine to the Renaissance*. Berkeley CA: University of California Press.
Mumford, Lewis (1970). *Pentagon of Power*. New York: Harcourt Brace Jovanovich.
Nagel, Thomas (1986). *View from Nowhere*. New York: Oxford University Press.
O'Donnell, James J. (1985). *Augustine*. Boston MA: Twayne Publishers.
Pippert, Wesley (1989). *An Ethics of News: A Reporter's Search for Truth*. Washington DC: Georgetown University Press.
Ricoeur, Paul (1973). 'Ethics and culture: Habermas and Gadamer in dialogue', *Philosophy Today*, 17, 2/4 (Summer 1973).

Rorty, Richard (1979). *Philosophy and the Mirror of Nature*. Princeton NJ: Princeton University Press.

Russell, Bertrand (1912). 'Truth and falsehood', *Problems of Philosophy*. London: Oxford University Press.

Rousseau, Jean Jacques (1961). *Emile*. London: Dent.

Settle, Glenn (1994). 'Faith, hope and charity: rhetoric as aletheiac act in On Christian Doctrine', *Journal of Communication and Religion*, 17:2 (September 1994).

Sullivan, Dale L. (1992). 'Kairos and the rhetoric of belief', *Quarterly Journal of Speech*, 78.

Thoreau, Henry David (1975). 'March 31, 1837', in *Early Essays and Miscellanies*. Princeton NJ: Princeton University Press.

Wiener, Jon (1989). 'Talcott Parson's role: Bringing Nazi sympathizers to the US', *The Nation*, 6 March 1989.

Wuthnow, Robert (1987). *Meaning and Moral Order*. Berkeley CA: University of California Press.

5

Democratization of communication as a social movement process

ROBERT A. WHITE

One of the central themes of communication policy discussion in recent years is how to enable citizens to have greater control over the processes of public communication. In the 'information society', the ability to define what is information and shape the flows of information becomes a form of political self-determination. The central issue in the debates of the New World Information and Communication Order has been the democratization of communication so that the public is guaranteed the exercise of the right to communicate.

Surprisingly, however, the democratization of communication remains a very poorly defined and explained process. The impression is given, however, that it is sufficient to draw up idealistic plans, perhaps with some consultation, and then find a friendly government executive to implement them. Indeed, media reform movements in Europe, Latin America, the US and other parts of the world have used this rationalistic design approach with few positive results. There is often little understanding of the social mobilization change in public attitudes that the democratization of communication implies.

It is the thesis of this chapter that the kind of social changes that are implied in the democratization of communication are best explained in terms of the process of social movements. That is, contemporary theory of social movements not only explains the conditions under which democratization of communication is likely to occur and develop, but will provide a much more complete and internally consistent understanding of the dimensions of the democratization of communication.

There are a number of basic points of intersection between social movement theory and what is commonly understood by the democratization of communication:

(1) Social movements are a communication pattern which emerges 'outside' and in opposition to the existing institutional, hierarchical (non-democratic) structure of communications in a society.

(2) Social movements, in order to strengthen identification and loyalty, tend to introduce and legitimate an alternative pattern of communication which, relative to the dominant pattern, insists that all members have a right to obtain and *make communicative inputs when they wish, that members may participate* in all phases of the collective communication decision-making process, that members may engage in 'horizontal' communication between individuals and groups without being vetted by authorities, that communication be *dialogical* in the sense that members have a right to reply and expect a direct reply.

(3) Social movements tend to renovate and democratize virtually all aspects of the communication process: the definition of what communication means; the definition of what social sectors and social actors may participate in the public communication process; the employment of new media technology and the democratization of existing technology; the redefinition of 'media professionalism' and the training of professionals; the development of new codes of ethics and new values guiding public policy, etc.

(4) What is central to the democratization of communication, most social movements insist, is that members – ordinary 'citizens' – should participate in the administration, policy-making and government of public communication.

(5) 'Epochal social movements', those social movements that introduce a major socio-cultural shift in civilizations also tend to introduce a radically different normative theory of communication and a new culture of public communication.

A conception of the democratization of communication

Descriptively, democratic communication refers to an institutional organization of public communication which attempts to guarantee the right of all individuals and subcultures to participate in the construction of the public cultural truth. Public cultural truth is the dominant consensus about what is true and what is the meaning of the history of the group or society at any given moment of time. For example, although at one time many may have accepted that human slavery may be a 'true', coherent interpretation of the social order and the basis for a national social project, today this would not be accepted as true in most societies. What is considered the public cultural truth is continually shifting, and communicators, from rhetoricians to public advocates, are extremely important in articulating and developing symbolic language to express what a people thinks is true. Nevertheless, the basis of the public cultural truth is the subcultures and, ultimately, the definition of reality of each person.

This conception of public communication rests upon a definition of communication as a process of negotiated convergence of meaning in which two or more persons, or at a societal level various subcultures, begin with their own definitions of the situation, but on the basis of a mutually involving action (co-operation or conflict), gradually create a new set of meanings which may incorporate something of the individual meanings but are unlike any single definition of meaning which existed at the beginning of the convergence process (Rogers and Kincaid, 1981). In so far as all of the parties in the process have participated in the construction of meaning and have contributed something of their own definition of the situation, each of the parties may recognize something of their 'identity' in the common meaning and there is a 'sharing' of meaning. But it seems more accurate to describe communication not as a 'sharing of meaning' but rather as a process of 'interactive construction of the meaning of a situation'. Communication is the process of continually constructing a new culture. It is 'public' in the sense that all of those who are involved with the process are aware of the steps in the construction of the meaning (what was agreed on, what was rejected, what was disagreed about, etc.) This brings out the fact that any given definition of the meaning of the situation is continually in the process of construction and change and that no single definition of the situation is completely 'shared' or 'identified with' by everybody in the process.

The institutional dimensions of the democratization of communication

Although the process of the democratization of communication will vary immensely from one context or level of concreteness to another, there are a series of problematics that are unavoidable in all institutionalized participation in the construction of public cultural truth. The fact that these problematics are continually present in most discussions of the democratization of communication, confirms their centrality.

1. The right to access to the process of constructing the public cultural truth

The most fundamental dimension of the democratization of communication is to guarantee all the information which is necessary for the basic human needs of education, health, personal development, occupations, and for significant participation in local or national public decisions. In most cases, this is not just a matter of information not being available, but also of it not being available in a form which is usable; the information is either unrelated to information needs or is presented to individuals lacking the socio-economic conditions to utilize it. The problem of the information rich and the information poor is characteristic not only of different social status groups within a national society, but also of national societies at the international level.

This responds to what is often referred to as *'the right to information'*. There is, however, in this insistence on the right to information, a lingering radical libertarianism which sees information as a help to individual goals and individual personal development. The image of this kind of society is the aggregate of separated individuals each responding not to any public definition of the common good, but rather to the interior voice of conscience. The common good is not any deeply shared meaning or value but the sum of unrelated personal goods. Individuals may agree to help each other achieve their personal goals – a kind of social contract – but this does not imply that either of the parties shares the same values or same goals. In fact, each may reject the conscience-based values of the other, but still agree to help each other achieve their personal goals.

It seems better, in this case, to insist on the fundamental right to be part of the process of societal interaction that is seeking to establish some degree of consensus about what is the public cultural truth. This is better expressed as the right to *'com-municate'*, the right to be part of an ongoing, dialogical conversation or debate by the person-in-community. This assumes that the person does not exist except in communication and community. It implies that the person must have access to information flows that are the resources for the construction of public cultural truth, but also that every person has the right to make an active contribution as an individual and as a member of a subculture to the formation of a public cultural truth. This means that as a result of the communication process, every person should be able to recognize something of his or her identity in the given historical moment of the public cultural truth.

For this reason, one is inclined to agree with Desmond Fisher that the *right to communication is a more fundamental right than the right to information, freedom of speech or other derivative rights.* While the right to communicate is so basic to human existence that it must always be without qualification, the derivative rights have conditions and limits (Fisher and Harms, 1982).

2. The expression of the right to communicate at the institutional, structural, societal level

The right to expression of one's personal creativity in the shaping of public cultural truth and the right to communicate are usually imaged and defined in terms of the individual within an interpersonal dialogue. When the right to communicate is described at the level of larger, more complex social groupings it becomes vague or to the point of disappearance or is very subtly suppressed under traditional ideologies that fear the wild, unruly voice of direct democracy in the mob. Freedom of speech becomes freedom of the press, that is, the freedom to be the proprietor of the press with the intervening caveat that only property owners have the right to vote or intervene in community affairs. Only those who have professional training or a well-developed sense of 'news values' are considered capable of responsible communication; public communication is a profession that should, like the profession of medicine, carefully exclude the layperson and the charlatan from practice in order to protect the public. Only those who have recognized talent as artists and 'stars' should be involved in public communication. Those who have invested their money in the infrastructure of public media have the right to determine how this money is spent 'responsibly'.

Democratization suggests that communication systems should be reorganized to permit all sectors of a population to *contribute to the pool of information* that provides the basis for local or national decision-making and the basis for the allocation of resources in society. All sectors of a population should have the opportunity to contribute to the formation of the national cultures that define their social values. All of the public should have access to the tools of media production and to technical help for making their own programming. Audiences should have the opportunity *collectively* to criticize, analyse, and participate in the communication process.

We anticipate a central part of our argument here by insisting that the fundamental obstacle to democratic participation lies in the adoption, as the basis of public communication – and conceptions of societal organization – of the modernization model of society organized around the formal bureaucracy. The rethinking of public communication in terms of the social logic of movement and empowerment suddenly opens the social imagination to the possibility of democratization at the macro, societal level.

3. Participation in the administrative and policy decision-making regarding the public mass media as the right of the citizen

It is taken for granted in most media systems that the public may, at best, be consulted in policy and administrative matters, but that the media have to run as an efficient business. You appoint managers to run a media corporation and that manager is accountable to the public in some way but is left free to run the corporation in an efficient manner on a day-to-day basis. The principle of efficiency in administration is only the tip of the iceberg, so to speak, because behind this is the efficiency of technology, the efficiency of economic planning, the efficiency of the political economy of the nation, etc. A little reflection suggests that the criteria of good functioning of virtually all media systems is the logic of instrumental efficiency that stems from the dualistic rationalism of the Enlightenment. Either an action is the most efficient way to carry out a particular action or it is irrational. This instrumental rationality runs through all policy and administrative decision-making: the decisions on what kind of programming will be made, policy and decisions on how the media system will relate to the socio-economic development of the nation; solving clashes of production values with public values on issues such as violence, pornography or censorship; the measurement of success in terms of quantitative size of the audience, etc. Behind this rationalism there lurk all sorts of attempts at hegemonic control of the masses, but the legitimizing face is that of rationalistic instrumental efficiency.

The logic of the public is different in that it is built around the logic of trying to make sense out of life, trying to affirm one's identities, trying to hold together the communal context of household and community of reference, getting snatches of pleasure, trying to discover oneself in free time, etc. This logic is much more holistic, multi-faceted, aimless and 'illogical' from the point of view of instrumental rationality. This is what some have called the logic of the popular which asserts itself more in resistance, in carnival, in ridicule and in cynicism than in the steady march of rationalistic planning. Democratization of communication struggles to open a space of legitimacy for this logic of the popular which is, in fact, much more multi-class than we might think.

Rationalistic administration, planning and policy in the area of media must deal with this popular logic in some way because this is its public. One of the major points of meeting and struggle is in the formation of genre. A particular genre of media is 'successful' with the public if it is able to pick up and incorporate the discourse and rhythm of life of a particular audience group and to maintain a steady struggle with that audience group over the control of the genre. The audience group, for example women who follow day-time soap opera, are never content with the genre because it never expresses their identity. The 'media' are always trying to use the sense of the identity of the audience to get to their goals of instrumental efficiency, for example, to increase the audience size or rationalize the budgets while the audience is always fighting back by deserting to alternative programming, mounting campaigns against particular writers, etc.

One 'out' for the almost defeated audience is to begin to create their own entertainment with their own version of the genre which truly represents their identity – their own music, their own fiction drama, etc. This 'audience creativity' reveals *the tip of the iceberg of potential democratization*. Democratization would open the door to far more audience creativity and to far more of the popular logic in the administration, planning and policy of media. The result would be a media in which the public can discover something of their identities.

4. The combination of property definitions, public legislation and public policy tradition which encourages a space of cultural dialogue among a broad diversity of interests

The movements toward the democratization of media have generally placed great emphasis on the separation of the media from one source of control: economic, political, religious, social class, cultural movements, etc. The mode of financing is crucial in this regard. One measure is to remove the media from dependency on capitalistic investment and market controls; another is to remove media from direct government control, control of political parties or the control (production or property) of churches. Each of these interest sectors operates in terms of a relatively neutral public forum where ascriptive power is removed – for example, the freedom of the economic market, parliamentary debate, the area of the sacred, etc. Yet each of these public fora requires a certain discourse or 'currency' that tends to exclude. Thus many societies have developed a combination of fora which act as mutual 'checks' on each other or a way of balancing diversity in the orientation of the public media: a mixed property system, market mechanisms with regulatory legislation and citizen supervisory boards, etc. Often these combinations still represent the interests and hegemonic cultural frame of an élite alliance.

What has not been obvious, however, is that efforts to attain a balance of power (or, a measure of social, contributive–distributive justice) in cultural representation cannot remain at the level of *political-economic* combinations. Rather, the guiding logic must move at the cultural level. That is, the founding conception of history, a cultural 'myth', of the civilization or the national versions of this civilization must be sufficiently open and self-critical so as to legitimate the cultural capital of *all cultural actors* and not permit any cultural actor which would introduce exclusive cultural norms. At the same time it must be a sufficiently strong conception of history or myth so as to attract a consensus regarding its maintenance as it continues to absorb an increasing number of diverse cultural actors.

This kind of conception of history encourages and legitimates the material expression of this openness in the kind of political-economic combinations discussed above but also continually and resolutely attacks the tendency toward rigid concentration of social power which is the constant tendency in the political-economic sphere. It is this dialectic between the cultural myth and the political-economic expressions that is encouraged in the social organization of the movement.

5. A new public philosophy of communication
The public philosophy of communication refers to the cultural values which structure all communication in a society at all levels, from the system of mass communication to dyadic interpersonal communication, and form the core of institutional roles and organization regarding communication in a society. Values are the concept of the good, and in this case we are speaking of the concept of good communication which every person has.

When democratic communication becomes the dominant public

philosophy it is embedded in a culture at the level of what Victor Turner refers to as the 'root paradigm', that is, the organizing principle of the world view, the ethos and the mythic conception of history which provides the basis for the collective existence of a society. Once the values of democratic communication are the organizing logic of all social institutions, providing the 'script', so to speak, for social roles of family, government, work, recreation, these values become the pervasive atmosphere of social relations.

A culture begins to exist precisely when there are mechanisms for passing values from one generation to another. Hence the development of institutions for socialization and formal education in democratic communication are central. Much of this is done through the informal culture of educational institutions, the ritual celebrations of a society, and the religious belief system that gives democratic communication a grounding in the sacred and the metaphysical.

6. A new normative theory

Normative theories are systematic presentations and justifications of the social institutions that are proposed as means to ensure the social living out of these values of the public culture of communication. For example, Milton, Cato's Letters, John Stuart Mill, Hamilton in the Zenger case all enunciated systematic explanations that came to form the libertarian normative theory of communication and public media. All public philosophies and normative theories of communication deal, in some way, with the procedures for establishing public cultural truth in a society and, therefore, deal with the fundamental issues of what is true and how a society can maintain the dimension of truth in its culture and communication (Altschull, 1990).

Most attempts to reformulate normative theories of communication are carried out in the name of democratization, and they do represent a degree of democratization. But most of them take a fatal turn that leads them back into exclusivism. For example, Milton argued in favour of free speech, even erroneous speech, because that would be the best step toward establishing the acceptable public cultural truth. But as a Puritan he was quite intolerant of Catholicism in Britain (Altschull, 1990). Virtually all other normative theories have, in their turn, become exclusive in some way.

The basic characteristics of social movements

We would define social movements as a social organization of leaders and followers to change power relations in the social system of which they are a part (Morris and Muller, 1992; Zald and McCarthy, 1987). We would include in this definition that the protagonists and members of the movement are those who are excluded in some degree from significant participation and influence on the collective decision-making of the hegemonic coalition of the system. Being marginal in some degree to the central communication and exchange networks, the potential protagonists of the movement have access to resources and information only through those individuals and groups that are more central in the networks. Thus, in terms of power relations, they are in a dependency relationship to the leaders of the hegemonic coalition. This usually means that the cultural capital, that is, the symbolic identification, of the protagonists is of lower value. Nevertheless some degree of equilibrium is maintained because the dependent groups (women, peasants, racial or religious minorities) maintain a somewhat separate sub-system with emotionally satisfying relations and a capacity simply to 'ignore in silence' the depreciative definitions of the group in the view of the dominant ideology. Most important is the need for the subordinate group to obtain, through individual client relations with more powerful people, the resources needed to resolve problems and attain minimal goals within the expectations of their subculture. A social movement begins when the marginal groups cut their ties of dependency through more central exchange and information networks and begin to mobilize an independent base of resources in order to gain more direct access and stronger influence in the collective decision making structure. Often this can mean a fundamental change in the bases of cultural power so that the cultural capital of the subordinate groups has much more validity.

It is important to note that social movements remain 'social' as long as they choose to operate through the developing of new symbolic and communication strategies and confront other sectors of society at the cultural level. The moment that a social movement opts for strategies of political, economic or military confrontation, it ceases to be a social movement and becomes a coercive campaign.

A social movement occurs in a context of general socio-cultural and political-economic change which causes a *relative* decline in the social power of the dependent groups, especially a greater depreciation of the symbolic identity. The dependent groups find the ordinary channels for obtaining help to solve problems gradually being closed off. Often, in the general socio-economic shift, other groups, which are points of reference to judge one's general welfare, are enjoying marked improvement of their status. What is often most painful for the groups whose power is declining is that it is becoming increasingly difficult to get access to *significant information* to solve fundamental problems in life. For example, in a general process of modernization the cultural identity of rural lower status groups (peasants) is seen not as worthless but as a positive obstacle. In the distribution of resources, the peasants are increasingly excluded from national planning. In this context, it becomes increasingly difficult to make sense out of one's situation and often, in the increased psychological stress, symptoms of mental illness and increased internal conflict begin to show up. Most often, however, the emotional energy to 'tear loose' from dependency relations and seek one's own resources comes only in a painfully sharp experience of powerlessness and frustration. This 'precipitating factor' dramatizes the dehumanization of the group and brings a profound emotional awareness that 'the situation is intolerable'.

Bringing marginal constituents into a network of access to significant information
An alternative network of communication, independent from the dependency relationships, begins to emerge at the moment that people of the marginal sector realize that they can obtain resources to solve crisis problems through their contacts with the centres of hegemonic control. At this point, the horizontal communication patterns that did not exist because peers often cannot help, because peers might be competitors in relations with patrons or because the horizontal relations are punished, become important. The processes necessary to build up an alternative structure of resources are all logical steps toward the democratization of communication.

It is in the interest of the leaders of movements to reach out to the largest possible number of dissident marginal adherents as possible and build solid loyalty to the alternative network because *one of the major power resources that these marginal dissident groups have is 'numbers'*. Thus, one of the key characteristics of movements among marginal dissidents is the *relationship of interdependence*, especially regarding information and communication.

Also, one of the major immediate resources that leaders can offer to potential adherents is *significant alternative information about the causes and solutions of the problems*. The nature of hierarchical dependency relationships to hegemonic centres of control is that they 'strain out' significant information because individuals of lower status identity are defined as having inferior 'cultural capital' and having nothing to offer except their obedience and submission. Moreover, to provide significant information, that is, information important for power, control and decision-making, would be a 'threat' to the existing relations of power. On the other hand, leaders in movements build up a 'culture of service', especially collecting information from within the network or from outside and *translating this information into a language understandable to the constituents*.

At the outset, the communication networks are largely oral and very localized – the interconnection of different small focal points of action. One of the single most important communicative transformations is that the depreciative self-identity of marginal groups that has built up in dependency relations with hegemonic control centres begins simply to disappear because it is no longer necessary to 'grovel' and 'feign incompetency' to get resources. On the contrary, it is in the interest of all members of the movement to assert the most positive interpretation of their identity and cultural capital in order to sustain confidence in their independence and build courage for encounters with more powerful sectors. One of the most characteristic aspects of movements is the *inversion* of self-depreciation into self-aggrandizement. For example, peasant movements invariably recast their image as the 'soul of the nation unaffected by foreign values', the 'preservers of the true solid values' of this culture, 'the land and the worker as the real base of any economy', etc. Racial and gender movements point out their biological superiority. Youth-based movements define themselves as those with the 'new realism', the hope of the future, etc.

This definition of the cultural capital has various significant implications for communication. Firstly, these new identity definitions become the *unifying symbols* and *a new common language cutting across barriers of locality, religion, family ties and other traditional divisions*. All people capable of sharing in this identity can enter into this language, and movements have a way of making these identity symbols as inclusive as possible. Secondly, the positive identification makes the movement attractive to all marginal sectors. Thirdly, movements have a way of building their own hegemonic alliances in the sense that they gradually build more and broader identity symbols by incorporating elements of many other movements and sectors. For example, peasant movements may begin to deal with modernizing sectors by arguing that the conception of development of this movement is, in fact, a contribution to modernizing, but 'authentic, integral modernization' or 'modernization with justice', or 'modernization led by the people'. Fourthly, the new positive identity symbols become the basis for a dramatizing rhetoric that projects the image of the dissident group into the public cultural sphere and attempts to get this cultural capital accepted within the synthesis of common symbols of the nation. Thus, a nation built around European Catholicism might gradually incorporate African Islamic cultural symbols as an integral part of the public cultural sphere. And, in so far as once-marginal groups gain more powerful access to the public sphere, they gain access to more significant information.

As Melucci points out, one of the essential strategies of movements is to project one's symbols into the public cultural arena and by that very action to contest and to delegitimate the existing symbols of hegemonic control (Melucci, 1989).

Building a participatory culture

Because numbers are a major source of power for marginal movements, the movement cannot afford to alienate anyone. One can observe in meetings, especially at the initial stage of a movement, the tendency for 'endless' discussion in which everybody has a right to voice their opinion and everybody is encouraged to speak out about doubts or disagreements in the meeting. Every point of view is taken into consideration, and, in the final decision, pains are taken to make it evident that the ideas of each person are part of the plan decided upon. Given the deep smouldering resentment over relations of dependency, for the leaders to take decisions without consulting their followers would be explosive. The more that the leadership can devise 'procedures' for participating, the greater the solid loyalty of the followers.

Indeed, the 'procedures which guarantee participation' become one of the most crucial identity symbols of these movements. Procedures of participation both dramatize to members that they have a 'guaranteed' *right* to participate and serve to distinguish this lower-status, marginal movement from the more authoritarian hegemonic identity symbols. Members look upon the procedures of democratic participation as part of the cultural capital of these movements making them superior to other less democratic organizations. Participation thus becomes part of the *'myth'* of the movement, portending the kind of communication and the kind of reformed society that they promise to bring into existence.

Participation also works its way into the fundamental ethic of communication. Just as lying is generally understood to be non-communication, so also non-participatory and non-dialogical communication is non-communication. 'Dialogue', the back-and-forth exchange that leads toward sharing meaning, gradually becomes the definition of the constituent nature of communication.

In a sense the agreed-upon 'procedures' for participation and the 'unifying symbols' of the movement are a form of incipient communication technology. That is, procedures are a rational codification of the values of the participatory culture and imply a reflexive 'normative theory' of communication. Procedures set the stage for participatory media in at least three ways: procedures guarantee that members will maintain control over all aspects of a communication process leading toward collective decisions, including decisions about the procedures; procedures ensure that leaders must be responsive to members and must reveal to members their actions at every stage; procedures insist that the final decisions be symbols of the unity of the group and that every member can recognize his or her identity in these decisions.

'Media' begin to be important in a movement when the movement grows beyond easy face-to-face communication and leadership cannot be immediately present to the members in every necessary communication. But media are also initiated when the membership enters into any form of communication beyond the casual interpersonal conversation. Examples of this are 'rhetorical dramatization' in meetings to motivate members to action; the use of narrative styles to recount, orally, the history of grievances and the history of heroes; the creation of symbols of the movement, such as heroes who have sacrificed themselves; the singing, joking and entertainment at meetings and among members; the creation of the language of cultural capital. Each of these artifices represents a *genre* of 'formal' communication which carries the meaning of participation, consensus, consultation of membership, articulation of

felt desires, building solidarity. The very process of trying to make decisions in meetings and articulate what people are thinking represents a 'participatory medium'. Almost always, the end result of this discussion is codified in a 'text' that remains as a symbol of participation to be interpreted endlessly by members on other occasions. Media are, thus, not just to 'transport' information, but to articulate and define cultural self-identity and create cultural capital.

Virtually all movements tend to take the communication technology which is most available to them – often, new technology which is not yet completely controlled by the hegemonic alliance – and 'embed' the technology within the communication procedures that have been created to guarantee participation and 'grass roots' control within the movement. Thus the press or radio are not inherently participatory, but become so because the procedures, genres, symbols and languages carry an inherently participatory logic.

Most movements have no difficulty in getting people to create 'participatory programming' which is of great interest not only within the movement but also within the public cultural sphere 'outside' the movement. The programming is of interest to the movement because it is based on a language which is understandable, articulating, congratulatory (positive images), mythical (indicates the society the movement wants to build) and carrying significant 'empowering' information. But the programming is usually of general interest because movements often develop considerable skill in devising symbols and communication strategies which are attractive, inclusive, dramatic and entertaining to the general public. Even when members of the public don't agree with the goals of the movement, the rhetoric captures attention and wins some loyalty.

The role of movement alliances in extending participatory communication networks

Movements of lower-status, dependent and marginal groups often do not have much experience in the use of communication technology. But movements are adept at making alliances with other dissident groups that bring to the enlarged movement highly skilled technical capacities. Paulo Freire, like so many other urban, technically trained people, was attracted to popular movements in Latin America because his search for cultural authenticity made him an outcast in his own social class and helped him to find his 'true home' in these movements. Freire and others learned the participatory philosophy from the movements, but once he had absorbed the values he was able to translate this into a method of popular education, a method of popular communication, a new conception of people's media, and a philosophy of popular culture.

Often these alliances link marginal groups into the heart of information exchange networks and give marginal movements access to both technical and *strategic* information. The dissident urban-technical sectors, outcast and punished by the hegemonic alliances, 'need' the popular base both to discover a new legitimation for themselves (authentic indigenous people of the country), but also because they represent the power of numbers. Thus these alliances constitute a form of communication channel that is creating in the movement significant, strategic and empowering information that further enhances the cultural capital of the movement.

Meeting the problem of size in mass media with the principle of federation

Communication at the small-group, face-to-face level is generally democratic to a considerable degree. The issue is one of maintaining participation and procedures of control when large-scale mass media are involved and administrative problems grow. At this stage, many would say that the limits of democratic communication have been reached and it is time to introduce the administrative structure of formal bureaucracy. At best, the public might indicate only indirectly what media policy is going to be and then leave the actual administration to paid professionals. Impersonal, professional bureaucracy takes over the decision-making and it is assumed the logic of professionalism will ensure that the decisions are for the public good.

Social movements, however, will not generally accept this solution because it implies the end of the movement and the end of democratic participation in the movement. If participation was precisely the reason for starting the movement in the first place (such symbolism is part of every movement), then this principle can never be sacrificed. Some way must be found to combine the participatory principle and efficient administration.

This logic of the movement leads toward the principle of federation and of subsidiarity. That is, the movement as a whole is very careful to maintain the purity of its culture and its symbols and the coordinating unity is maintained largely at the level of overarching symbols and *voluntary identification with these symbols*. The agreed policy is more at the level of symbolic identification. Administration is kept decentralized at the local level and local units have a great deal of autonomy in their administration. The local administrator is elected by the members, but once he is elected he is empowered to administer. A central procedure, however, is that one unit will not interfere with the financial or political affairs of another. Thus, the movement is held together largely by communication, especially communication in which the dramatic symbols and continual representation of the unifying 'myth' of the movement are kept before the members. Since the unifying symbols represent the identity of the members, the members want to be continually involved in the communicative process of defining policy and the central unifying symbols. Governing officers of federations are often elected because people can identify with their values and the way their policies articulate the feelings of the people.

Thus, democratic administration of mass media is achieved by maintaining a multiplication of local media units which have considerable local control. The local media unit will reflect the local culture, interests and creativity. The people will be able to develop the genres they most identify with. 'Quality' and 'professionalism' in this kind of media rest on the ability to articulate authentically the feelings of the local people and enable them to create something they can truly identify with. From time to time, a local production may have much wider articulating power and it may be reproduced by other local units by popular desire. This organization of media and this kind of communication are never really 'mass' in the sense that a central administrator is making the decisions about what will be presented without consultation with the constituent members.

Constituting the media as a public cultural sphere in which all cultural identities are represented
As was noted above, one of the most challenging problems in the democratization of communication is to prevent one cultural discourse from becoming highly dominant so that it depreciates and excludes the validity of other cultural capitals. To the extent that one cultural discourse becomes hegemonic, it becomes so closely linked with the sense of national unity and destiny (what can be called the cultural myth of the society) that this discourse becomes an ideology

that uses the people of the nation for its own sectorial purposes in the name of the national unity. The classical example of this is the nationalistic war of expansion that one social class promotes for its own aggrandizement and that other social classes have to pay for and die for. In this case the media, too, become instruments of class ideology.

One response to this problem has been to separate media from control by economic, political, religious and other sectorial interests and to define the media as a public cultural sphere in which all have equal right of access.

I would argue that a more radical response to this tendency toward the concentration of cultural capital and the formation of ideologies lies precisely in the processes of social movements that promote the formation of alternative cultural capital, the capacity to project this into the public cultural sphere and the capacity aggressively to negotiate the power of cultural symbols in the public cultural sphere. Regulation of proprietary control, for example, may keep one sector from dominating and it may open up the media to other sectors. But nothing will happen unless other sectors mobilize culturally and take advantage of the more open public cultural sphere. Indeed, one could argue that the demand that the media be made an open forum has come from the pressure of dissident cultural movements.

As we noted above, one of the first moves of an alternative network of social and communication relations that is becoming a movement is to assert and develop its cultural validity. This process is begun in the rhetoric of mobilizing a constituency and building loyal solidarity with the movement. Very quickly, however, virtually all movements develop their own media which can collect and present to the constituents the 'texts' which are the outcome of local meetings, speeches, demonstrations and cultural actions. Presenting these 'texts' to the constituents enables them to appropriate more consciously and reflexively the new cultural identities that are emerging in the process of cultural action. This method of reflecting their reality back to the members of a movement was codified by Paulo Freire in his education for liberation (Freire, 1987). This Freirian approach to 'conscientiza-tion', 'consciousness raising' and 'empowerment' has been picked up by movements around the world and applied to all sorts of small media: video and audiovisuals, role playing and 'street theatre', people's radio and television, and even by such simple methods as graffiti and posters. It is important to note, however, that Freire only codified a process that is an ordinary dimension of all social movements and that 'conscientization' does not exist outside social movements. What this has brought out, however, is the importance for the definition of cultural capital of small media that the group can control and use continually to evaluate and reformulate texts that articulate cultural identities.

A social movement is, by definition, an alternative social organization that seeks change in society, and no social movement is content with defining its own cultural identity. The process of internal mobilization develops a persuasive rhetorical discourse that tells its constituents, 'Your identity is not only valuable, but it is also worth defending and expanding.' Thus every subculture carries the imprint of an apologetic and vigorous offence with other cultural fronts (González, 1994).[1]

In order to protect itself, every movement must project its identity into the public sphere. One element is simply to persuade other subcultures and movements to let one's own have the right to exist. Another element is to recognize that other cultural movements have valuable elements for coping with the environment and that one might learn from them. This entails cultural 'borrowing' as long as it is clear that this borrowing is used and interpreted in terms of one's own cultural identity. A final element is the recognition that other cultural movements are individually and collectively important for the existence of one's own culture and that the public sphere must open itself to many different symbolic dimensions. The history of virtually all social movements shows that they move from an initial stage of assertive defensiveness that gives way to interchange, tolerance and support of other cultures once it is clear that trying to build a cultural ghetto is a self-defeating goal.

It is important to note, however, that a movement may give up trying resolve the problem of its isolation at the cultural level, but may short-circuit this process and seek to transform the cultural sphere through political, economic or military force. Here the logic is that cultural accommodation is never attained by communication or dialogue. I would argue that at this point communication ceases within the social organization that calls itself a movement, and a movement is transformed into a political-economic and, eventually, a police or military campaign.

One might rightly ask if there is any role for a deliberate policy of democratization if the creation of a public cultural sphere is an 'automatic' result of social movements. It is true that this does not square well with a functionalist social engineering approach, but it is crucial simply to recognize the priority of the cultural and to provide a space for it to operate. One example of this process of recognition is in what I have called 'public cultural rituals' (White, 1990). Such rituals are a forum which are, by definition 'popular' and playful, removed from work and politics, removed from dogmatic purism to reinforce boundaries and where symbols of community come to the fore. Here all of the cultural fronts come together and symbolically engage in jousting with each other simply by the power of their cultural capital. Some of the great moments of the media are these public cultural rituals. What these rituals do is to call different cultural fronts out of their isolation, encourage them to dramatize themselves to the maximum, look seriously at other cultures and find some common bonds under overarching symbols.

Embedding the democratization of communication into the paradigmatic logic of a culture

As we saw above, the democratization of communication is not brought about simply by passing certain legislation or introducing a new policy. The values of participatory communication must become deeply a part of cultural identities so that, in every context, people automatically organize social relations in a participatory and dialogical fashion. The only way that a public policy or a particular form of legislation toward democratization will be respected and enforced is if everyone who has some bearing on democratic communication procedures believes that this is *vital* for human existence. But how are such deeply felt convictions developed in a culture? Again, I think that a *culture* of democratic communication emerges in the context of a social movement.

Social movements, by definition, are attempting to create an alternative communication pattern and one in which the members have greater access to information. Its 'weapons' are symbolic strategies. As was noted above, a movement needs to encourage democratic participation and dialogical communication in order to build loyalty. Very often (I hesitate to say always) participatory communication becomes defined into the cultural *identity of the movement and becomes part of the 'text' which is held up to the people.* At this point, participatory, dialogical communication is no longer simply a useful strategy but becomes a *symbol* detached from its origins, a value in itself and part of the collective identity of the movement. Participation is used to differentiate this movement from the hegemonic coalition and members come to understand that respecting democratic communication is a condition of being a member of the movement. Thus, participatory communication becomes embedded into the fundamental objectives of social transformation held out by the movement. If participatory communication becomes deeply embedded into the root paradigm of the culture of the movement, then this is developed as a key element in the cultural capital and is presented in all cultural negotiation as the most valuable thing that this movement, as a cultural front, has to offer other social sectors. Thus, democratic communication becomes negotiated into the common, public cultural sphere that all subcultures must recognize in some way if they are be part of that culture.[2]

Developing a normative theory of democratic communication

Once a movement begins to consider democratic communication as an identifying symbol and this becomes part of the 'texts' of the movement, it is often only a short step toward developing a theory to explain 'why' participatory, dialogical communication is a preferred value. I would argue that the greatest threat to developing democratic communication as an embedded public philosophy and as a highly developed normative theory, is when the movement finds the opportunity to become part of the hegemonic coalition. The longer a movement remains a marginal 'prophetic' voice, the more likely it is that it will have time to develop a social ethic of democratic communication.

In societies where no particular movement is able to develop such powerful symbolic hegemony so as to invalidate other cultural capitals and the society is characterized by continued fluidity of movements, the more likely it is that the symbols of democratic, participatory and dialogical communication will be an enduring and operative part of the public cultural sphere that all groups must respect.

What this essay has attempted to develop is the view that democratization is an emerging structural process. It is based on the premise of Marx's famous dictum, that all people are free in their decisions, but they may not be able freely to affect the *conditions* in which they are forced to make those decisions. If democratization of communication is to take place, it is important to become aware of the structural processes, the processes underlying the conditions, and to some extent to be able to seize the moment to create an opening for democratization. By being part of this process of structural change, we may hope to respond to the possibilities for democratization when they are in some degree open to us.

Notes

[1] Cf. González (1994) chapter 2, describing the dynamics of cultural movements and cultural fronts.
[2] An example of this process is the case of Mediterranean Christianity. During its existence as a subaltern marginal subculture in the pre-Constantinian Roman empire, Christianity was a democratizing subculture. But over more than a millennium of absolute cultural hegemony, Mediterranean Christianity became Roman Catholicism that drained out all sense of cultural tolerance. As long as Protestantism thought of itself as 'reformed Roman Catholicism' and maintained cultural hegemony in a given country, it also had little time for dialogue. But once Roman Catholicism and Christian groups became once again marginal cultures – movements rather than churches – they rediscovered the values of dialogue and participation.

References

Altschull, J. Herbert (1990). *From Milton to McLuhan: The Ideas Behind American Journalism*. New York: Longman.

Fisher, Desmond and L. S. Harms (eds.) (1982). *The Right to Communicate: A New Human Right*. Dublin: Boole Press.

Freire, Paulo (1987). *A Pedagogy of Liberation*. New York: Continuum.

González, Jorge (1994). *Más (+) cultura(s): Ensayos sobre realidades plurales*. México: Consejo Nacional para la Cultura y Artes.

Mayer N. Zald and John D. McCarthy (1987). *Social Movements in an Organisational Society: Collected Essays*. New Brunswick: Transaction Publishers.

Melucci, Alberto (1989). *Nomads of the Present: Social Movements and Individual Needs in Contemporary Society*, eds. John Keane and Paul Mier. London: Century Hutchinson.

Morris, Aldon D. and Carol McClurg Muller (eds.) (1992). *Frontiers in Social Movement Theory*. New Haven: Yale University Press.

Rogers, Everett M. and D. Lawrence Kincaid (1981). *Communication Networks: Toward a New Paradigm for Research*. New York: The Free Press of Macmillan Publishing.

White, Robert A. (1990). 'Cultural analysis in communication for development – the role of cultural dramaturgy in the creation of a public cultural sphere', in *Development*, 2, 23–32.

6

The journalist: A walking paradox

KAARLE NORDENSTRENG

> Although millions of people work in communication in one way or another, special attention is rightly devoted to journalists. They have not only an important social function, but their potential capacity to influence and even to shape ideas and opinions . . . makes journalism both a profession and a mission. This is particularly important since public opinion is dependent more than ever on those who supply objective, truthful and unbiased news and information; the news gatherer and news disseminator are essential to the workings of any democratic system.
>
> (*Many Voices, One World*, 1980: 233)

The MacBride Commission paid due attention to journalism as one of the central communication problems of our time. The Commission was very supportive of this kind of mass communication, with Seán MacBride himself promoting the idea of special protection for journalists. The above quote, from the beginning of chapter 5 ('Rights and responsibilities of journalists') of the MacBride Report, highlights the approach to journalism which was so typical of the 1970s – journalism basically seen as a positive factor for democracy, in need of further professionalization.

With hindsight, one could say that the Commission, in line with the dominant mood of the day, was uncritically naïve – or romantic – about journalism and journalists. It failed to see the contradictions involved in the very nature of journalism 'as a profession and a mission', beginning with the anti-democratic tendencies associated with any strong profession. Hence its recommendations concerning the journalistic profession – for enhancing its standing in society, elevating its educational level, raising its professional standards and responsibility, and ensuring its accountability towards the public –

remained largely wishful thinking, reconfirming good old ideas and making little difference in actual practice.

Today, when CNN and the information superhighway are flourishing, it is fashionable to be critical about journalism to the point of prophesying 'the end of journalism' (cf. Katz, 1992). True, there are trends away from the great and honourable tradition of print-dominated journalism, understood as a pillar of democracy, towards a multimedia information and entertainment machine, understood as a pillar of the market economy, with few traditional (full-time) journalists – indeed few traditional (all-round) mass media.

Yet this chapter does not subscribe to such a vision of future media with little or no role left for journalism. The present author happens to believe that the mass media in general, and journalism in particular, continue to be of vital importance to societies both in the North and the South – not least with regard to their democratization. But this belief is held with mixed feelings, far from the affirmative and celebratory positions of the 1970s and 1980s, since today one cannot fail to see that journalism is full of contradictions – indeed, it appears as a dilemma and a paradox waiting to be deconstructed.

Dimensions of the paradox

The paradox has many faces but there are four aspects or dimensions that emerge as the most obvious: accuracy, rapidity, seriousness and autonomy. They are summarized here without elaboration.

First, *accuracy*. Truth in the sense of factual accuracy is no doubt the most sacred belief held among journalists worldwide, with the related demand for professional ethics to correct mistakes. A recent survey of some thirty codes of journalistic ethics in the European region confirmed the centrality of truthfulness: nine codes out of ten include this provision which stands at the top of a long list of aspects covered by the codes (Laitila, 1995).

But the fact-oriented concept of truth carries with it a bias in the same way as logical positivism is limited in discovering social reality: while faithful to the surface it misses the deeper structures beneath. Moreover, the tendency to separate facts from opinions leads to journalistic strategies to 'objectify morality by transforming moral claims into empirical claims' as well as to privatize and narrativize morality, thus 'dissolving moral discourse into an empty rhetoric of blame and praise, celebration and condemnation' (Glasser and Ettema, 1994: 338).

It is indeed paradoxical that such dissolving of morality and overlooking essentiality is based on the ideal of journalism as an unbiased window, mirror, lens or whatever form of 'glassy' substance which is supposed honestly and faithfully to portray reality. The lesson taught by this first dimension of the paradoxical nature of journalism is that we have to question even the most fundamental dogma of the profession – truth seeking – because the way it has been conceived and practised in journalism serves as a deceptive filtering device preventing as much as helping the truth being discovered. Actually this is not a new dilemma but something that has been debated for quite a while, especially in relation to the objectivity of journalism. Yet, a truly philosphical treatment of the theories of truth has remained superficial as shown by the unproblematicized way in which truth in general and accuracy in particular is still conceived in journalism.

Second, *rapidity*. Speed, with truth, is among the most central constituents of journalism – after all, the very name of the profession comes from a daily, if not more instant, reporting of reality. A mild version of this paradox is reflected in the old ethical rules for checking the facts and correcting the errors; it was understood that reporting too rapidly may be detrimental to accuracy. Moreover, it was recognized half a century ago by the Hutchins Commission that journalists should place daily facts in their proper context or else the audience is unable to form an adequate picture of the world.

Lately, however, the paradox of rapidity has been dramatically increased with the development of the electronic media and real time journalism. The new media environment has an overflow of information – the kind of information which tells a lot about instant details but very little about fundamental developments. In fact, real time journalism is a contradiction in terms; live and full coverage of the events themselves can indeed be seen as the end of journalism, or on the contrary, as an invitation to truly interpretative journalism complementing the simple transmission of events.

Third, *seriousness*. Reporting accurately and rapidly has implied a choice of topics with significance of one sort or another – politics, the economy, etc. Historically, newspapers started with this serious material, which was later complemented by sports, entertainment, and so on. Much of the lighter part of the menu – cartoons and popular columns – in fact served as the digestive to the serious main course of journalism. A mild version of this paradox was understood for a long time in the form of a contradiction between objectively significant news and subjectively interesting news, the latter serving as a vehicle to convey the former to the reader's mind. The underlying paradigm was the same as in education: the audience was supposed to learn something, to be enlightened.

This traditional concept of journalism is challenged by another development of the new media environment, taking place parallel to real time journalism and absorbing serious journalism by entertainment. It does not just consist of more human interest material to ensure that the rest of the news gets home, and it is not just more scandals and other sensational material with the purpose of selling the paper or the channel – ultimately for the advertisers – but it is a mixture of facts and fiction, information and entertainment. 'Infotainment' is taking over journalism, with serial shows participating in daily politics – real politicians playing with actors – and radio talk shows breaching other holy conventions of journalistic genres.

This paradox is not only about the moral dilemma of a serious journalist being co-opted by the entertainment industry. At issue is a more fundamental question: journalism has often failed to get its message across, whereas entertainment seems to be doing well not only at keeping audiences but also at telling serious stories. Cinema and fiction literature are well known for their performance in this respect, but there has always been a demarcation line between journalism and other cultural genres. Now this line seems to be fading away.

This development is not just to be regretted as a victory for light entertainment, immoral commercialism, etc. The clue to the paradox is that a non-serious, narrative mode may in many cases lead to better enlightenment of the audience than traditional serious journalism does. As with the first dimension of accuracy, we are here faced with a kind of tragedy whereby the best of professional standards may prevent rather than facilitate the intended outcome. Fiction and opinion, rather than fact and neutrality, seem often to produce the best portrayal of reality. In this sense the return of the storyteller in an electronic form is welcome, and the real kind of doubtful escapism should be seen in conventional, serious journalism. The challenge is enormous – not only to journalists but also to their educators.

Fourth, *autonomy*. Journalists, like the mass media in general, need a degree of independence from socio-political and economic forces in order to reflect reality free of various power constellations. This is another basic dogma of the profession, widely held internationally as shown by a survey of first-year students of journalism in twenty-two countries, with striking similarities emerging precisely in terms of 'a desire for the independence and autonomy of journalism' (Splichal and Sparks, 1994: 179).

The dogma is turned into a paradox by the fact that autonomy will easily lead journalists into a self-centred 'fortress journalism', alienated from the people whom it is supposed to serve. As a matter of fact, the very nature of professionalism tends to isolate any professionals from the people; the professionals necessarily becoming more or less technocrats. This is a general paradox of professionalism, but it is particularly acute in journalism – something that concerns most sensitive reflections about socio-political reality.

By and large, the paradox of journalism can be viewed as a great irony, as suggested by Ettema and Glasser (1994). As the muck-rakers of the Progressive Era in the US failed to stimulate political action and thus created an irony of voter apathy at a time of aggressive journalistic enlightenment, the investigative journalism of our time, as well as all the other hyperinformation, seems to foster apathy and alienation among citizens rather than the civic virtues prescribed by theories of democracy. It is indeed a deep irony of journalism if the citizen is lost in an information society; if 'the audience will merely watch and . . . witness the finale of its own annihilation as a public' (p. 27).

Such a development is still a thought experiment rather than a universally established fact. Some observers, for example Galtung (1994), suggest quite a different future, with civil society – rather than the state and capital – dominating social development. The present author shares Galtung's optimism, which however does not do away with the paradox of journalism characterized above. On the contrary, a prospect for a future with genuine democracy will place the theory and practice of existing journalism under increasing challenge.

Media in democracy

Three key players in media–society relations are the journalist (media), the politician (government) and the citizen (people). The dynamic of this triangle is illustrated in the following figure where arrows show in which direction the influence goes – according to the ideal theory of democracy, on the one hand, and according to the real practice of democracy, on the other. In democracy theory, media are

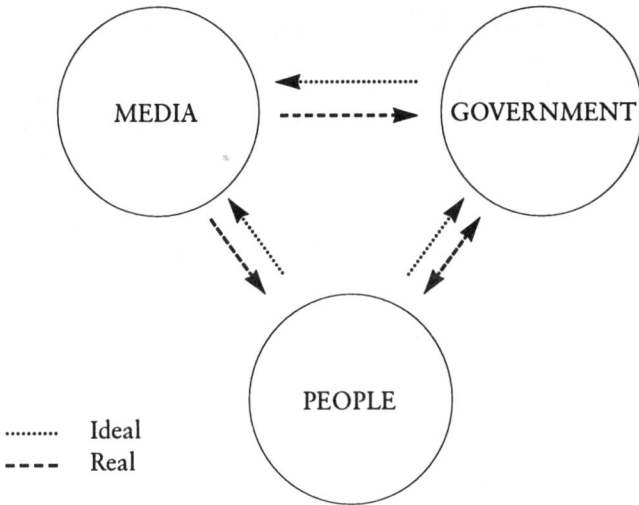

Figure 1: Media in democracy

supposed to be in the service of the people – as are politicians running the government for the people. Therefore an ideal relationship of determination is from the people to the media, both directly and via government.

In real life, however, media exercise a strong influence on both people and government, thus occupying a master's place rather than that of a servant in the power game. Ideal and real relationships are reversed, the people being a *target* of influence, instead of a *source* of influence, except for elections where people periodically choose the politicians (in the USA only about half of the electorate even does this).

Although such a textbook illustration simplifies the relationships and generalizes the variations, its main point remains valid: democracy is not living up to its ideals, and the media and journalists constitute a central part of the problem. The triangular relationships also point the way to democratization: the media should be brought closer to the people and there should be at least a reciprocal flow of influence between the two – regardless of relations to the government. Accordingly, there is need for a shift of power from the media to the people, as suggested by Hamelink (1994) in his draft for a People's Communication Charter.

Conceptually, the shift calls for a determined move away from the traditional notion of a self-centred profession – fortress journalism – towards a position whereby the owner of the right to information is the citizen instead of the media. It is time to release the concept of freedom of information from the hostage it has been taken by media proprietors and to return it where it was originally placed in Article 19 of the Universal Declaration of Human Rights – to 'everyone', i.e. the citizen. In other words, the main function of the media in the context of democracy is service to the people, rather than an abstract mission to seek for truth. Freedom in this democratic design belongs to citizens rather than to media.

Such a shift can be criticized for advocating populism, but all the same it is more in line with the theory of democracy than is the fortress journalism concept. As noted above, the performance of journalism in western democracies, beginning with the USA, has been far from perfect despite all the professionalism and its ethical intentions; perhaps 'poor' would be the proper description. Thus the shift is motivated not only by a theoretical argument but by a very practical argument: the media are failing to keep people informed and democratically engaged. Moreover, the print media are losing readers, especially young readers, and there is a growing concern among proprietors to look for new strategies in industrial survival.

One response to this situation is a new movement of 'public journalism' in the USA (Rosen, 1994). Also known as 'civic journalism' it calls for a community connectedness, which was part and parcel of the early forms of the press but was largely lost in the process of modernization. It is ironical indeed that today the same commercial corporate forces that for decades drove the media to be more and more commercial are open-minded and even supportive of initiatives to mobilize the public and to reinvigorate the democratic process. It happens to be in their current commercial interest to join the intellectual movement back to people and to grass-roots democracy. No doubt this is part of a more fundamental trend in the secularized and materialized West towards social and communitarian values – a trend countered by increasing selfishness and hard-line politics.

Public journalism is an innovative exercise in the theory and practice of journalism, fostered by Jay Rosen's 'Project on Public Life and the Press', based at New York University. It has successfully demonstrated in several communities throughout the USA how press and also electronic media can be brought back to people's agenda and turned into an exciting instrument of political participation instead of perpetuating alienation and disintegration. For journalists it requires a professional approach which is quite different from the conventional role: a neutral information transmitter is supposed to turn into a moderator of grass-roots politics.

It is likely that a public or civic journalism movement will emerge in Europe and the rest of the North as well, because the social and political niche is there. As for the South, developmental journalism has already provided the conceptual and practical response to challenges there. As a matter of fact, public journalism is bringing to western journalism basically the same paradigm as has been cultivated in developing countries for decades – and opposed by many western media lobbies as something detrimental to freedom (like the rest of the New World Information and Communication Order).

If the conceptual shift from the media to the public, highlighted by the public journalism movement, is singled out as perhaps the most vital trend in contemporary journalism, the second most vital trend would be self-regulation. The latter is as old as the codes of ethics and courts of honour for journalism, which were first created by the emerging profession between the two World Wars. But self-regulation is gaining particular importance at the present time due to increasing pressures which are directed at the media – from the political élites (the state in Galtung's triangle), from market forces (capital), and from the public at large (civil society).

Faced with these pressures, which ultimately boil down to the growing role played by the media in social and political processes in the postmodern era, professionals resort to self-regulation as a defensive strategy particularly in order to avoid governmental intervention. And many progressive intellectuals join to support self-regulation of the media against state or market regulation, as a means of protecting freedom and democracy in society – just as happened fifty years ago with the Hutchins Commission.

Self-regulation with professional ethics as its central element can be seen to be contradictory to the popular shift discussed above – after all, the idea of self-regulation is based on an autonomous profession which sets its own rules and thus is quite a long way from the idea of service-oriented public journalism. On the other hand self-regulation also means recognition of social responsibility, i.e. accountability to society and its various strata, which conceptually reconnects media to the people. After all, the very idea of accountability implies criteria from the outside, and the question is what are the social values and norms to be voluntarily adopted by the self-regulating media. You simply cannot escape society, except by an anti-intellectual strategy of denial, and therefore a serious attempt to promote self-regulation is bound to consider not just media autonomy but also the social values being served.

In this respect self-regulation fits well within the overall theory of democracy. But it has to be taken seriously and exercised actively or else it remains the kind of alibi which has been the function of most journalistic codes and media councils so far. Today it no longer pays to deny outside intervention by referring to the mere existence of the more or less dead letters of a code or other purely formal instruments of self-regulation; the instruments in place must be real and effective. There may not be a need for other forms of self-regulation than those already known – promoted by the MacBride Report – but they must be actively developed and kept alive.

It is obvious that journalists to a large extent, and also media proprietors to a certain extent, are ready to enter a new era without sticking to established principles and practices. Precisely what will follow is hard to foresee, but the trends outlined above are likely to have a central place. A textbook example of innovations based on existing tradition is provided by Galtung and Vincent (1992) in their sets of ten proposals for peace-oriented news media, development-oriented news media and environment-oriented news media.

At this stage it is worth reviewing the historical passage so far travelled in international efforts to define and improve the profession (based on the author's chapter in Nordenstreng and Topuz, 1989).

Historical highlights

Efforts to improve the status of the journalist and to articulate his or her rights and responsibilities are not of recent origin. As a matter of fact, these efforts have as long a history as the professional and trade-union organizations in this field – the first national unions of journalists having been established in the 1880s and the first international congress of 'press people' convened in 1894. The first high season in this respect was from the mid-1920s to the mid-1930s when the International Labour Organization (ILO) and the League of Nations gave rise to the first international organization of working journalists. It was at this time that such questions as special cards for journalists and an international court of honour were elaborated.

These efforts were paralysed by the rise of Fascism and the coming war, but a new drive among journalists began during the Second World War, leading to the establishment of the present International Organization of Journalists (IOJ) in 1946. The immediate post-war period can be characterized as the second high season with a broad consensus among professionals from East and West to pursue freedom and responsibility of journalists as well as trade-unionism among them. Typical of this situation was the highest possible status granted to the IOJ in the United Nations Conference on Freedom of Information in Geneva in spring 1948.

The Cold War brought this season to a halt before it produced anything tangible beyond statements and resolutions – which as such were quite thoughtful and promising. Unfortunately the international movement of journalists was split in much the same way as was the world at large. However, from the mid-1950s onwards there were significant attempts to promote professional interests on a broad international basis, notably through so-called World Meetings of Journalists (Helsinki 1956, Baden 1960, Mediterranean 1961). After these, regional organizations of journalists took shape in Africa, the Arab world and Latin America.

These developments were significant in political and regional terms, but they did not produce much in terms of a major global improvement in the status of journalists or in the understanding of their rights and responsibilities. It is also to be noted that the Cold War paralysed much of the UN and Unesco activities in this area until the late 1960s – the first and last major initiative which Unesco undertook during this period being the attempt in 1948–9 to set up an International Institute of the Press and Information.

It was only in the early 1970s – along with East–West détente and the consolidation of the Non-Aligned Movement – that a new historical momentum was reached for true international collaboration in the interest of the profession. At this stage Unesco came to play an active role, among other things by promoting ethical principles. These efforts may not have been very innovative – much of the substance was simply a repetition of what was said during the earlier high seasons – but none the less they were significant, both politically and professionally. This is particularly so because of what appears to be a kind of ecumenical approach covering not only various geopolitical orientations but also workers and employers alike.

This can be regarded as the third high season, lasting until the late 1980s, with landmarks such as the codes of ethics by the regional organizations in Arab, Latin American and ASEAN regions as well as a joint venture by the Finnish and Austrian unions of journalists, the International Federation of Journalists (IFJ) and the IOJ in the context of the Conference on Security and Co-operation in Europe. Of particular importance at this historical stage was the MacBride Commission which played a catalytic role in focusing the attention of both Unesco and the professional journalists' organizations on issues of the status, rights and responsibilities of journalists worldwide.

One of the steps taken by Unesco at the time was to invite the international and regional organizations of working journalists to hold a consultative meeting at its headquarters. The meeting in April 1978 decided that those involved would set up a system of regular consultation and of eventual joint action particularly in the areas of professional ethics and solidarity. This led to the so-called Consultative Club which has since met in Mexico City (1980), Baghdad (1982), Prague and Paris (1983), Geneva (1985), Brussels and Sofia (1986), Cairo and Tampere (1987), Prague (1988), Mexico City (1989) and The Hague (1990).

The most far-reaching joint venture by the Consultative Club to date is a document entitled 'International Principles of Professional Ethics in Journalism' (for the text see Traber and Nordenstreng, 1992). As the present author put it in a brochure promoting the document, the Principles 'constitute the first time ever that the profession of journalism has manifested itself in a universal declaration of ethics'. The document, adopted in November 1983, was an outcome of a collective effort of the Club, which represents the overwhelming majority of organized journalists in the world. Not less significant, all continents and geopolitical regions were covered, which means that behind the document were quite different ideological and philosophical orientations extending from Communists to Christian Democrats. Yet it is obvious that the document enjoyed little or no support at the political extremes – particularly among those who sympathize with 'tyrannic regimes'. The document itself had a political orientation which might simply be called democratic. It was a manifestation of the same line of universal values as advocated by the Mass Media Declaration of Unesco in 1978.

Such a dedication to the values and principles of the international community was a significant step for a profession with a strong tradition to remain independent and, in particular, free from government interference. After all, in practice the international community is made up of governments, although, in theory, the concept ultimately refers to the peoples of the world. Yet it should be noted that nothing in the document suggests that the professionals concerned would welcome governments to assume a greater role in mass communication. It simply means that the profession itself is dedicated to the same universal values and principles that are reflected in the UN system and in international law.

By and large, the document prescribes journalism as a socially committed profession. The commitment originates from the people's right to acquire a truthful picture of objective reality, on the one hand, and from the universal values of humanism on the other. The commitment to truth is, in principle, the same as that held within the libertarian mainstream of journalism, although there are obvious differences between traditions as to the nature of truth. But the commitment to universal values as established by the international community means a significant departure from the typical Western tradition and a move towards the notion of professionalism as generally understood in the then socialist and developing countries.

Accordingly, 'a true journalist', as defined by the document, is not neutral with regard to the universal values of 'peace, democracy, human rights, social progress and national liberation'. Neither is a journalist neutral with regard to violations of humanity such as 'justification for, or incitement to, wars of aggression and arms race, especially in nuclear weapons, and all other forms of violence, hatred or discrimination, especially racialism and apartheid, oppression by tyrannic regimes, colonialism and neo-colonialism.'

Consequently, the document stands for a concept of professionalism which, while building on the established traditions of journalism, commits the journalist to certain universal values. This does not mean that, objectively speaking, the profession would be less independent than under a doctrine which has made the libertarian notion of freedom its value foundation. Journalism is always bound to be dependent on certain social interests and values, whether openly recognized or accepted as a hidden ideology. In this respect the document plays an important function as an instrument to stimulate critical appraisal of the profession itself.

On the other hand, the new 'committed' professional ethics

appears to be a less remarkable leap forward than was suggested above. After all, the journalist does no more than become openly committed to the values which constitute the foundation of international law and order. If this seems to be a radical step, it only goes to show how poorly universal values have been recognized, often due to the dominance of parochial values which stand in opposition to those held by the international community.

Thus, the new professional ethics in journalism did not bring any particular 'politization' into the field of information; it only provided a safeguard against policies which depart from the universally recognized values of peace, democracy, etc. It goes without saying that journalism is and will continue to be a highly political field – overtly or covertly. In such a situation, any choice of professional ethics represents a direct or indirect political position. The question is not which is political and which is apolitical; the question is what is the political orientation being advocated. In this respect, the new professional ethics had as 'impartial' a foundation as can be imagined: the universal values of the international community.

The 1983 Principles serve as a reminder that professional doctrines in journalism were in transition – after decades of intellectual stagnation. This was true of all the three 'worlds' – East, West and South. The latest reading of the Club's position to the same effect is to be found in the Final Report of the International Symposium on the Mass Media Declaration of Unesco, submitted to the Unesco Secretariat in May 1988:

> We wish to reiterate the principal view that the operation of the mass media should be determined primarily by the practice of professional journalism in the public interest without undue government or commercial influence. What we stand for is professionalism supported by the idea of a free and responsible press.
>
> We acknowledge the fact that the role played by information and communication in national as well as international spheres has become more and more prominent during the past decade, with a growing responsibility being placed upon the mass media and journalists. This calls, increasingly, for professional autonomy of journalists as well as a measure of public accountability. (Nordenstreng, 1993: 105)

This statement, like the 1983 Principles, serves as a textbook example of the democratic shift discussed above: a professional ideology of fortress journalism is yielding to a new doctrine taking people more seriously and admitting self-regulation as a mechanism of accountability. On the other hand, the statement (particularly its beginning where professionalism is defined to be 'primary') still carries a legacy of fortress journalism – not surprising given the source of the statement: a collective voice of journalists worldwide.

All and all, the Consultative Club and its statements such as the one quoted above suggest an optimistic reading of how professionalism has developed over the decades – indeed over a full century. Despite all the divisions in the world, a universal force of democratic journalism has emerged. Certainly, it has not penetrated all media in all countries; what is at issue is not a total trend but rather a significant trace of a trend. Moreover, even if penetrated, it does not guarantee that actual media performance is changed accordingly. Naturally doctrines and statements do not change the world, but they are necessary vehicles in mobilizing journalists for the cause of democracy.

Prospects ahead

If historical development has produced three high seasons of promoting the status and doctrines of journalism up to the 1980s, one is led to ask: 'Isn't there room for another high season just now – after the collapse of Communism in Eastern Europe, beyond the Cold War?' The answer is No, or at least not Yes.

It is indeed paradoxical that at the time when the Berlin Wall came down, the universal movement of journalists got paralysed. This was partly caused by the turmoil which entered the IOJ as most of its East European member organizations (beginning with the one at its seat, Prague) went through a total political change. But it was also caused by the fact that the IFL (with its political environment in Brussels which was unchanged after the Cold War) was no longer interested in talking to other regions at the Consultative Club but instead only on its own ground. In other words, the Cold War legacy took its toll by turning earlier, more or less equal partners into an impossible equation: IOJ on the defensive, IFJ on the offensive – with the regional federations in Africa, Asia and Latin America torn by their own internal conflicts. No doubt it is a complicated story with many factors involved, but it remains a sad fact that the Club has been totally dormant since 1990.

One contributing factor to the paralysis of a universal movement of journalists is Unesco, which effectively isolated the IOJ and took the

IFL as the only international partner among journalist organizations. As the IOJ had been *primus inter pares* in the Consultative Club, Unesco distanced itself from this forum of co-operation. Instead Unesco promoted its own regional conferences of independent media, including media owners. In this new constellation, the IFL was brought to the same table with newspaper and broadcasting proprietors, promoting media freedom around the world – but no longer in the name of working journalists alone as had been the case at the Club table.

Accordingly, the earlier 'ecumenical journalist' movement got no support in the new political environment, and it was replaced by a more or less vindictive 'free media' movement. However, this page of history is not yet finished. It may well be that a new and more balanced approach will follow – both in international journalist movements and Unesco – but it will hardly be so sweeping and deep as to give rise to another historical high season.

Apart from politics, the prospects ahead are made gloomy by the developments in media ownership and structures. Media concentration and transnationalization, with the ever greater role of market forces, leads to journalists and their profession being under heavier pressure. One aspect of this development is the fact that fewer and fewer creative media workers are employed full time and more and more are working freelance. This weakens trade unions and reduces their potential for promoting the status of the profession, both nationally and internationally.

But even this gloomy trend has its countertrends. Hard times also stimulate a fighting spirit; trade unions and professional associations have by no means ceased to be influential in the media field. At the international level, the IFJ not only sits with the employers in exporting freedom to the East and South, but it keeps fighting the employers and their organizations in matters of media concentration, copyright, etc. Moreover, the IFJ has become a leading force in the fight against racism and xenophobia in the media – as invited by the 1983 Principles.

Still, keeping in mind the dimensions of the paradox discussed at the beginning of this chapter, one cannot but conclude that the profession is in disarray and in a state of confusion. Although we should not become doomsday prophets and propagate the fashionable vision of 'the end of journalism', the prospects ahead are far from clear and obvious.

Uncertainty invites vigilance. No simple problems, no easy solutions. A lot of challenges.

References

Ettema, James S. and Theodore L. Glasser (1994) 'The irony in – and of – journalism: A case study in the moral language of liberal democracy', *Journal of Communication*, 44(2), 5–28.

Galtung, Johan (1994). 'State, Capital and the Civil Society', keynote paper at the 6th MacBride Round Table in Honolulu.

Galtung, Johan and Richard Vincent (1992). *Global Glasnost. Toward a New World Information and Communication Order?* Cresskill, N.J.: Hampton Press.

Glasser, Theodore L. and James S. Ettema (1994). 'The language of news and the end of morality', *Argumentation*, 8, 337–44.

Hamelink, Cees (1994). *Trends in World Communication. On Disempowerment and Self-empowerment*. Penang: Southbound and Third World Network.

Katz, Elihu (1992). 'The end of journalism? Notes on watching the war', *Journal of Communication*, 42 (3), 5–13.

Laitila, Tiina (1995). 'The Journalistic Codes of Ethics in Europe', report for the WAPC Conference in Helsinki, 1 June 1995.

MacBride, Seán (1980). *Many Voices, One World*. London: Kogan Page; New York: Unipub; Paris: Unesco.

Nordenstreng, Kaarle (1993). 'The story and lesson of a symposium', in George Gerbner, Hamid Mowlana and Kaarle Nordenstreng (eds.), *The Global Media Debate: Its Rise, Fall and Renewal*. Norwood, N.J.: Ablex Publishing Corporation, 99–107.

Nordenstreng, Kaarle and Hifzi Topuz (eds.) (1989). *Journalist: Status, Rights and Responsibilities*. Prague: International Organization of Journalists.

Rosen, Jay (1994). 'Making things more public: on the political responsibility of the media intellectual', *Critical Studies in Mass Communication*, 11, 362–88.

Splichal, Slavko and Colin Sparks (1994). *Journalists for the 21st Century*. Norwood, N.J.: Ablex Publishing Corporation.

Traber, Michael and Kaarle Nordenstreng (eds.) (1992). *Few Voices, Many Worlds*. London: World Association for Christian Communication.

7

Women and communications technology: What are the issues?

COLLEEN ROACH

> Go forward, not backwards,
> Seize time, seize training opportunities.
> Teach yourself. Set your own horizons.
> Take note. Take courage. Take care.
> Take no tricks of tradition,
> Which hide you, hold you down, keep you back,
> Take hold firmly of tools and technology,
> Take part fiercely in the future.
> Take stock of changing times.
> Take on a stake in training. Be all that you can
> . . . and all that you want.[1]

Communication technology is a vital aspect of the democratization of communications. This is evident, most recently, in the discourse accompanying the much-vaunted global design for information technology known as the Information Super-Highway (ISH) of the Clinton-Gore administration. At a conference in October 1992, entitled 'Media, Democracy and the Information Highway', the theme of democracy and new technology was a key element. Sponsored by the Freedom Forum Media Studies Center in New York, the various speakers, updating such classics as Alexis de Tocqueville's *Democracy in America*, cited the various ways that democracy is linked to the creation of a vast web of computer-related information services. For example, Alfred C. Sikes, former FCC (Federal Communications Commission) chairman, defined the 'concept of democratization' as 'ensuring broader dissemination, ultimately universal access to new technologies . . . Putting more information in the hands of people will provide the building blocks that will help our citizens become better-informed and better-involved people, who will be able to make better-educated decisions about their everyday lives and also about world events' (Freedom Forum Media Studies Center, 1993: 4). In a similar vein, John Carey, director of a telecommunications research and planning firm, noted that 'the focus of democracy is the individual citizen and it is with the citizen in mind that the information highways and byways must be developed', (Freedom Forum Media Studies Center, 1993: 8).

Given this obvious connection between democracy and information technology, highlighted by the discourse of the ISH, the question is: where do women fit into the picture? This chapter will examine the key issues involved in the question of 'women and communications technology', highlighting some of the various feminist and gender-focused perspectives involved.

Communications technology: Relationship to Technology, with a capital T, and the science question

In the 1970s, when gender-related aspects of communications became a topic of research in the field, concerns in the West centred on two major areas: *(1)* images of women in the media (mainly television, movies and advertisements) and the concerns of how these images were related to violence towards women, sexism and female stereotyping; and *(2)* the structural aspects of job-related discrimination towards women working in communications fields (primarily, although not exclusively, journalism) (Roach, 1994). However, although both of these areas continue to be well researched, since at least the mid-1980s a new area of concern has increasingly come to the fore: women and communications technology. This relatively new area of research builds upon and is intimately related to two larger issues of concern: Technology itself and the science question.

Although many people think of technology merely as machines, most writers who have thought about the subject to any great extent, see it as something much more all-encompassing than just physical objects. Jacques Ellul, for example, makes a distinction between 'la technique' and mere machines (technology strictly speaking), emphasizing that 'la technique' is a way of life that began in the Middle Ages, picked up steam during the North's industrial revolution, and was bequeathed to modern-day societies which deify means over ends and rationality/progress over more humane values (Ellul, 1964). Many other writers who have a less sweeping panorama of technology, nonetheless see it at the very least as being a set of social relations and specific values with social and ethical implications (Traber, 1986).

Feminist and women writers, building on this larger body of the critique of technology have engendered a new contribution to the literature, evident in recent writings.[2] These works have injected notions such as gender, power and patriarchy into their work. For example:

> When we talk about technology, we are not only talking about a product or process (a home computer or automated factory, for example) but also about a whole set of ideas and values that go into the design, making and use of such a product or process. Thus in the context of a world where men hold most of the powerful positions and control the use of resources, we understand technology as being imbued with essentially male centered values.   (Zmroczek *et al.*, 1987: 121)

Or:

> Technology is assumed to be one of the many patriarchal attempts to dominate nature and consequently, women . . . The result . . . is a socio-cultural system in which men have power, whereas women are excluded from power, and in this case from the design, production and use of technology.   (Frissen, 1992: 5)

Technology, with a capital T, in turn is related to another, larger body of critical work on the relationship between women and science. Sandra Harding, whose award-winning *The Science Question in Feminism* (Harding, 1986) is considered a classic, has been one of the pioneers in the critique of science. In a more recent work, she points out that recent feminist challenges of science and technology take place within an overall context of 'rising scepticism' about the benefits of science and technology. However, she sounds a note of caution by warning that 'these discussions also occur when intellectuals in the fields of science and technology are gaining more and more power in higher education and government' (Harding, 1991: 1).

This work is of particular interest because, unlike many Western feminists, Harding makes a deliberate attempt to connect her critique of science and technology to the concerns of countries in the South. In *Whose Science, Whose Knowledge?* she has a chapter specifically devoted to the ways in which the 'First' World both benefited from and curbed scientific discoveries in the 'Third' World.

In taking the position that the history of science is actually one of 'common histories, common destinies', Harding notes the following:

> To try to explain the rise of science and technology in the West without referring to the Third World histories with which that rise is causally linked (and vice-versa) can produce only partial and distorted accounts. Intentionality is not the issue here. For the purposes of a causal account, it is not relevant that many Western scientists and technological innovators have not intended to de-develop the Third World in order to develop their explanations or disseminate their innovations (though some have had just these intentions).
>
> From such a perspective, we should refer not to the development of Europe and the underdevelopment of the Third World but to the *over*development of Europe and *de*-development of the Third World.
>
> (Harding, 1991: 234)

Harding also takes note of Third World writers who have themselves challenged Western primacy in science and technology such as Susantha Goonatilake (Goonatilake, 1984). This author's work challenges the supposedly universalist Western assumption that the history of science in the West includes the history of *all* science. According to Harding, Goonatilake paints a gloomy picture of the continuing situation of Third World dependence on the West in matters of science and technology, which parallels their socio-economic dependence (Harding, 1991: 231).

As writers such as Harding and others indicate, there are two major aspects of the 'feminism and science' question: *(1)* women in science; and *(2)* the role of gender ideology in science.

The dearth of women in science is of major concern to all who have examined figures in this area. In a 1993 *Scientific American* article Margaret Holloway cited some US statistics which, she says, sound like a 'warped version of the Twelve Days of Christmas': 1 per cent of working environmental scientists, 2 per cent of mechanical engineers, 3 per cent of electrical engineers, 4 per cent of medical school department directors, 5 per cent of physics Ph.D.s, 6 of close to 300 tenured professors in the country's top ten mathematics departments, and so on (Holloway, 1993: 95–6).

Holloway's lengthy article uses dire terms to refer to the absence of women in science:

> Although their struggle to enter and to advance in this overwhelmingly male-dominated field parallels the struggles of women in other

professions, science seems a uniquely well-fortified bastion of sexism . . .
Despite speeches, panels and other efforts at consciousness-raising, women
remain dramatically absent from the membership of the informal
communities and clubs that constitute the scientific establishment. Only 1
per cent of the employed scientists and engineers in this country are female.
(Holloway, 1993: 94)

Another woman writer, based at a major American research
university, noted:

Yet at my university (which is all too typical of research universities) only 7
per cent of the tenure-track positions in the natural and mathematical
sciences are occupied by women. There are no female faculty members of
any rank in computer science, only one each in chemistry and geology, and
a grand total of two each in our large physics and mathematics
departments. (Koertge, N., 1994: A80)

Although it is beyond the scope of this chapter to examine in depth
the reasons why there are so few women in science, it is clear that a
starting point for analysis would have to be the gendered aspects of
the first division of labour. It is well known, for example, that
researchers in universities and private laboratories have very long
days, sometimes putting in 70–80 hours per week. As long as societies
are structured in such a way that women bear primary responsibility
for child care and labour in the home, they cannot possibly match the
time requirements met by their male counterparts.

Another aspect of the 'women in science' subject relates to the
absence of knowledge about the contributions of women in the
history of science. Like African-American and Third World writers,
women have attempted to 'resurrect' knowledge about the
contributions of women scientists, ranging from ancient figures to
contemporary researchers. Hypatia (c. 370–415), for example, was an
Egyptian mathematician, teacher and philosopher who was murdered
by a group of monks. Ethel Browne Harvey (1885–1965) was an
American biologist whose work in induction preceded by many years
that of Nobel laureates in the field, but who was never even made a
full professor at Princeton University, where she worked for twenty-
five years (Holloway, 1993: 99).

The role of gender ideology in science raises far more complex and
controversial issues. The different feminist critiques of the ideology of
science point out that this body of knowledge, in spite of its fabled
'objectivity' and 'neutrality' is also socially constructed, that is to say,

it is gendered knowledge. Some feminist scientists have pointed out that most modern-day science is the brainchild of Enlightenment notions of objectivity, rationality, and progress, which are held responsible for many of the ills of modern-day life: threats to the ecosystem, unchecked urbanization, and the threat of a nuclear holocaust. Feminist peace researchers have incorporated critical work from both science and technology into their far-reaching analysis of the causes of war (Roach, 1993).

The feminist philosophical critique of science was summarized by one writer in the following terms:

> Feminist philosophers look at science – and therefore at technology – as an historically specific social construct, where what are called 'facts' and 'proofs' may not necessarily represent universal truths or objectively determined laws of nature. This epistemological analysis asserts that there is a false universalism in claiming such objectivity, and in defining what is knowledge while at the same time denying any subjectivity to such definitions. What is called 'rational' thinking is one particular mode of thinking whose primacy has been aligned with European men, and used to discredit other approaches, specifically to accuse women and other non-dominant groups of being irrational and therefore not worthy of serious consideration. (Pritchard, 1993: 7)

Much of the philosophical basis for the feminist critique of the ideology of science is what has been termed 'feminist standpoint theory'. This theory is not unlike what some Third World writers refer to as 'situated knowledge' (Parajuli, 1991). One male writer summarizes this theory of knowledge in the following words: 'Feminist standpoint theory argues that race, class and gender are relevant to how one sees the world. Indeed, it goes further, by insisting that those situated at the margins are better positioned to understand the world than those in power' (Baskin, 1994: 13).

The stakes in science are obviously very high, given its role in modern-day technological societies. The 'power of progress' probably explains the depth of the controversy surrounding feminist questioning of the very bases of scientific enquiry. One obvious solution to both the gender-based critique of the ideology of science and the 'numbers' problem would be to change educational patterns and recruitment so that there are more women scientists in all countries of the world. However, even increasing the numbers of women in science has been subject to debate. A recent article, in the highly influential

U.S. Chronicle of Higher Education, illustrates how the two aspects of the question of 'feminism and science' have now become merged in polemics. Tellingly entitled 'Are Feminists Alienating Women from the Sciences?' this article, authored by a woman, argues that feminist questioning of scientific enquiry is actually keeping women out of science:

> A senior sociologist, highly regarded as a scholar and a founding member of the National Organization for Women, described to us the hostility of feminist graduate students towards courses on statistics and quantitative methods . . . We concluded that the ethos of contemporary women's studies not only would discourage young women from seeking a career in science, but also would make them feel morally and politically enlightened in remaining ignorant about science. (Koertge, 1994: A80)

The author concludes by attacking the feminist argument that ' "women's ways of knowing" are incompatible with the allegedly masculine substance and method of science. What young women really need is special encouragement and equal opportunity to learn science, not a feminist rationalization for failure' (Koertge, 1994: A80).

The so-called 'feminist rationalization for failure' is undoubtedly linked to what this author herself cites as one of the most profound attacks of feminist writing: the notion of scientific objectivity. In fact, scientific objectivity (which, as critical communication scholars have demonstrated, occupies a similar privileged pedestal in media practices) is definitely the deepest bedrock of the supposedly neutral spirit of science.

A much more interesting position that validates *both* scientific objectivity and a gendered interpretation thereof, is articulated by a male writer addressing this question. William Baskin argues that taking a feminist standpoint actually enhances the possibilities for scientific objectivity:

> . . . the scientific community is privileged in terms of race, class and gender. Yet, according to the traditional concept of objectivity, race, class and gender have no place in science. This means that the scientific community has no way of identifying (let alone eliminating) the racist, classist, and sexist values from which it benefits and which it helps to maintain. By recognizing that knowledge is situated, that every point of view is necessarily partial, and that this partiality must be made explicit,

feminist standpoint theory strengthens objectivity. Values which would have operated anyway behind the scientists' backs are laid out on the table where they can be dealt with.   (Baskin, 1994: 14)

Feminist and gender aspects of communications technology

What then are the aspects of communications technology questioned by feminist writers and researchers concerned with gender? In assessing the literature of women and communications technology (most of which is being produced in the West) it is very clear that current research reflects the many strands of thought in feminist literature itself. (It is increasingly commonplace, in fact, to read that one must speak of feminism*s* and not feminism in the singular.) Liesbet van Zoonen's analysis of the literature, although somewhat schematic, points to two major schools of thought: the 'liberal feminist'; and the 'radical feminist' (van Zoonen, 1992). The 'liberal feminist' viewpoint posits that the main issue for women in communications technology (computers, video cameras, VCRs, fax machines, etc.) is one of access. Women must have access to the power of technology, which in itself is viewed as primarily neutral. Proponents of this position would most likely join hands with the author cited above who argues that in science what women need most is 'equal opportunity' and not to question the gendered aspects of communications technology.

At the other end of the spectrum, one finds what van Zoonen labels the 'radical feminist' viewpoint. Proponents of this position argue that since science and technology are inherently masculine and patriarchal, women should turn their backs on communications technology and rely on women's more 'essentialist' nature of caring, and establishing connections with other human beings without the intermediary of machines such as computers. In this category of thought one would find writers referred to as 'ecofeminists' (e.g. Mies and Shiva, 1993) who stress the threats to human ecology posed by technology and science. Ecofeminists also emphasize that the traditional feminine values should combat patriarchal norms embedded in all forms of technology. Other writers in the 'radical' category would take the position, like women who reject traditional science courses, that 'wilful ignorance' is bliss, as evident in the following citation: 'I believe I am fortunate in not understanding it [effects of information technology]; I may embody the discontinuity that is going to allow technology and people to get along together'

(Davis, 1985: 168). Others still build upon feminist analysis of language (e.g. Spender, 1990) to stress that the very language of communications technology, particularly computers, is masculine and patriarchal. Women's exclusion from the construction of this language is but another example of how they are 'silenced' by the power inherent in technological construction.

Nonetheless, there are a range of positions and concerns articulated by women (again, primarily in the West) on the topic of information technology. To put it succinctly, what both the research literature and popular press indicate is that many women make use of information technology (particularly computers) but want to transform it to meet their own needs or to challenge its sexist assumptions and practices. Researchers have shown, for example, that although the telephone was originally mainly a tool of commerce, women's usage of this technology transformed it into a mainstay of communities and the social network (Moyal, 1992; Rakow, 1992).

Other feminists have built upon what Sherry Turkle has termed women's 'computational reticence' (Turkle, 1988), not to suggest that women opt out of the game, but rather that they *redesign* the game with new rules and less hierarchical modes of interaction. In the United States there are now a number of on-line services mainly designed and run by women, such as: Meta Network, ECHO (East Coast Hang Out), and Women's Wire (Kantrowitz, 1994: 42–3). Still others have stressed, as in the science debate, that if 'computer culture' is to change more women must enter the field. Men earning computer-science degrees in the United States outnumber women three to one, according the National Science Foundation (Kantrowitz, 1994: 39).

Women who fully participate in the computer age have also raised a number of issues and even legal challenges on the use of cyberspace, such as the following: on-line pornography and erotica, stalking and sexual harassment via E-mail, and sexually explicit information networks. Again, the point here is that women have not adopted the 'radical' stance of rejecting participation in computer networks, but have attempted to transform a reality summarized by one writer in the following terms: 'Cyberspace, it turns out, isn't much of an Eden after all. It's manned by just as many sexist ruts and gender conflicts as the Real World' (Kantrowitz, 1994: 36).

Even so, one must be mindful of the possibility that the debate on 'women and communications technology' as outlined thus far seems to be largely conducted by and of relevance to women in the West. This observation is supported by my own participation in a seminar on 'Women, Science and Technology', sponsored by Calcutta University and the US Educational Foundation in India (Roach, 1994). The seminar was very instructive for the purposes of this chapter because it demonstrated that the topic of 'women and communications technology' means different things in different parts of the world. It would not be at all surprising if the debate, particularly on the gendered ideology of science and technology, has little meaning for a country such as India. For the eminent Indian women scientists at this meeting, the key questions were clearly those of access and equal opportunity in the field. Even more to the point, of the various papers presented at the seminar on 'communications technology' it was clear that for the vast majority of participants this topic was related to images of women in traditional media technology, such as television and advertisements, and even literature. Given the fact that India has a vibrant women's movement, clearly in evidence at this seminar, it is therefore not inappropriate to question the degree to which other priorities condition women's concerns in other parts of the world.

Moreover, a brief look at some of the writings on communications technology by women from the Third World who have more activist (as opposed to academic) concerns, clearly shows that their priorities revolve around using the media as a form of 'alternative' communications (Ramirez, 1988; Viezzer, 1988; Mol, 1991). Such writers reflect the ethos promoted by Mike Traber, who has defined 'alternative communications' as 'alternative to what people receive from above. Alternative communication, it would seem, is the right way to respond to the communication challenge of our time' (Traber, 1986: 4).

Conclusion
There is every indication that, at least in the West, the 'communication and technology' question will continue to be a major concern for feminist researchers. It is also evident that the popular press has taken up interest in the issues involved in this question, as shown in a lengthy treatment of the subject published in *Newsweek* (16 May 1994) as its cover story, 'Men, Women and Computers'. However, judging from the literature on the Information Super-Highway, it is much less evident that the 'gender and communication

technology' issues have impacted upon the high levels of policy-making. Women must therefore make every effort to make known their views to the higher echelons of government, the corporate world, and any other decision-making body involved in setting priorities for the ISH.

However, women who are trying either to change the gendered nature of science, Technology, and communications technology, or improve their access to these fields, should not only attempt to affect those in high places. Rather, because of their position 'at the margins', so to speak, women are also uniquely positioned to struggle for those in the lower strata of society, whether from their own ranks or other subaltern groups. In Mike Traber's words:

> Women's position is similar to that of the poor, the manual labourers, the marginalised, the minorities, both ethnic and religious, the alienated youth, the lower castes, the people with dark skin . . . Herein, then, lies the true significance of tackling women's issues on communication. The adoption and acceptance of women's values in public communication could change our culture of silence . . . Their values will not only liberate women but also a host of other marginalised and oppressed groups.
>
> (Traber, 1991: 1)

Notes

[1] 'New Directions in Non-Formal Education, Papua New Guinea,' reprinted in 'Women and Technology', *WE CAN* (AR-WACC-PIC Women's Desk Quarterly), III(1), December 1991, p. 2.

[2] See, for example, Benston, 1983; Hynes, 1991; Lewis, 1987; Rothschild, 1983, 1988; Zimmerman, 1983.

References

Baskin, W. (1994). 'Feminism and Science': Working Papers of the Michael Harrington Center for Democratic Values and Social Change. Queens College, New York.

Benston, M. L. (1983). *The Technological Woman: Interfacing with Tomorrow*. New York: Praeger.

Davis, R. M. (1985). Comments. In B. R. Guile (ed.) *Information Technologies and Social Transformations*, pp. 168–9. Washington DC: National Academy Press.

Ellul, J. (1964). *The Technological Society*. New York: Knopf.

The Freedom Forum Media Studies Center (1993). *Media, Democracy and the Information Highway: A Conference Report on the Prospects for a National Information Service*. New York: Freedom Forum Media Studies Center.

Frissen, V. (1992). 'Trapped in electronic cages? Gender and new information technologies in the public and private domain: An overview of research.' *Media, Culture and Society* 14(1), 31–49.

Goonatilake, S. (1984). *Aborted Discovery: Science and Creativity in the Third World.* London: Zed Books.

Harding, S. (1986). *The Science Question in Feminism.* Ithaca: Cornell University Press.

Harding, S. (1991). *Whose Science? Whose Knowledge?* Ithaca: Cornell University Press.

Holloway, M. (1993). 'A lab of her own', *Scientific American,* November, pp. 94–103.

Hynes, H. P. (ed.) (1991). *Reconstructing Babylon: Essays on Women and Technology.* Bloomington: Indiana University Press.

Kantrowitz, B. (1994). 'Men, women and computers,' *Newsweek,* 16 May, pp. 36–43.

Koertge, N. (1994). 'Are feminists alienating women from the sciences?, *The Chronicle of Higher Education,* 14 September, p. A80.

Kramarae, C. (ed.) (1988). *Technology and Women's Voices.* New York: Routledge and Kegan Paul.

Lewis, L. H. (1987). *Women, Work and Technology.* Ann Arbor: University of Michigan Press.

Mies, M. and V. Shiva (1993). *Ecofeminism.* New Delhi: Kali for Women.

Mol, A. L. (1991). 'Experimenting with home language instruction for Moroccan women via interactive cable', *Media Development* XXXVIII (2), 28–30.

Moyal, A. (1992). 'The gendered use of the telephone: An Australian case study', *Media, Culture and Society,* 14(1), 51–72.

Parajuli, P. (1991). 'Power and knowledge in development discourse: New social movements and the state in India'. *International Social Science Journal,* XLIII(127), 173–90.

Pritchard, S. (1993). 'Feminist thought and the critique of information technology', *The Progressive Librarian* (New York), no. 8, pp. 1–9.

Rakow, L. F. (1992). *Gender on the Line: Women, the Telephone, and Community Life.* Urbana: University of Illinois Press.

Ramirez, M. M. (1986). 'Communication as if people matter: The challenge of alternative communications', in M. Traber (ed.) *The Myth of the Information Revolution,* pp. 99–116. Newbury Park, London, New Delhi: Sage.

Roach, C. (1993). 'Feminist peace researchers, culture and communications', in C. Roach (ed.), *Communication and Culture in War and Peace,* pp. 175–91. Newbury Park, London, New Delhi: Sage.

Roach, C. (1994) 'Women and communication technology: What are the issues?', keynote lecture at Seminar on Women, Science and Technology, Sponsored by the US Educational Foundation in India and the School of Women's Studies, University of Calcutta, Calcutta, India, 16 March.

Rothschild, J. (ed.) (1983). *Machina Ex Dea: Feminist Perspectives on Technology.* New York: Columbia Teachers College Press (Athene Series).

Rothschild, J. (1988). *Teaching Technology from a Feminist Perspective: A Practical Guide.* New York: Columbia Teachers College Press (Athene Series).

Spender, D. (1990). 'Extracts from man made language', in D. Cameron (ed.), *The Feminist Critique of Language: A Reader,* pp. 102–10. New York: Routledge.

Traber, M. (ed.) (1986). *The Myth of the Information Revolution: Social and Ethical Implications of Communications Technology.* Newbury Park, London, New Delhi: Sage.

Traber, M. (1991). 'Editorial', *Media Development,* XXXVIII (2).

Turkle, S. (1988). 'Computational reticence: Why women fear the intimate machine', in C. Kramarae (ed.), *Technology and Women's Voices,* pp. 41–61. New York: Routledge.

van Zoonen, L. (1992). 'Feminist theory and information technology', *Media, Culture and Society,* 14(1), 9–30.

Viezzer, M. (1986). 'Alternative communication for women's movement in Latin America', in M. Traber (ed.), *The Myth of the Information Revolution,* pp. 117–25. Newbury Park, London, New Delhi: Sage.

Zimmerman, Z. (ed.) (1983). *The Technological Woman: Interfacing with Tomorrow.* New York: Praeger.

Zmroczek, C., Henwood, F. and Wyatt, S. (1987). 'Women and technology', in G. Ashworth and L. Bonnerjea (eds.), *The Invisible Decade: Women and the UN Decade, 1976–1985.* London: Gower.

8

Traditional communication and democratization: Practical considerations

PRADIP N. THOMAS

> While we might occasionally use the term *traditional* to refer to a certain
> aspect or type of popular culture that comes into being through opposition
> to modernity, such words must be read in quotation marks . . . as formulas
> used for their functional value, to identify *phenomena*, not essences, that
> exist and that need to be given a name. (Canclini, 1993: 27)

Typically, the traditional and the modern occupy polar ends of the
continuum of civilization, development, worldview and 'structures of
feeling'. There is a wealth of associations, understandings and
assumptions related to the two concepts. Both have been used as
theoretical handles, as 'gatekeeping concepts', as 'metonyms and
surrogates' (Appadurai, 1986) to describe and account for profoundly
complex, polyvariant and multi-dimensional life processes. The
recourse to fixed, organic, dyadic constructs to make sense of life is a
peculiarly Occidental 'knowledge trait'. However, the assumed
universality of dualistic models of life based on such dyads as good
and evil, the modern and the traditional, etc., is immediately called
into question when confronted with the litmus test of lived
experiences from around the world. For there are continuities between
experiences and concepts as well as gaps, shades, blurrings and
breaks. Not that dyadic categorizations are alien to the Oriental
mind, but that these are, more often than not, used in a malleable,
flexible sense, subordinated to common sense and context and to an
'open' understanding of order.

The word 'traditional' has many different connotations, in the
sense that its meaning is derived in context. A traditional way of life
may denote a culture – nomadic, agricultural, etc. – that is rooted in a
material and normative order which is different from or only

marginally conversant with the gamut of modernity. The culture of the Amish in North America and the Ibans in Sarawak, among very many others, is often called traditional. But the word also refers to a culture-specific 'structure of feeling', our inherited understandings, popular memory and the means with which we negotiate identity, make sense of life, of place and space in the context of modernity. It is also often used to describe those communities which adhere to exclusivist, essentialist readings that are based more often than not on religion-derived understandings of order. Traditional ways of action are based on codes, 'traditions' which are encrypted in texts as well as in the collective memories of people. All people are socialized into traditions and these are transmitted in most societies by oral means from generation to generation. Tradition, therefore, needs to be seen both as a process and as the transmission of a specific symbolic discourse through given processes.

Traditional forms of communication are inherently symbolic in content and form. Metaphor, metonomy and the use of discontinuous analogies are some of the means by which symbolic meanings are communicated in tradition. Typically, traditional forms of communication are composite, in the sense that they are expressed through the dialectic of structure and anti-structure. Both resistance and accommodation can and do occur simultaneously in traditional communication and this is often achieved through recourse to role-reversals, the recombination of forms and the transformation of content. In this sense traditions are continually being reinvented. Traditional forms of communication are tremendously varied and include diverse forms of mediation – the formal as well as the informal. The formal includes all practices of communication that are organically related to the celebration of community. For example, those forms of communication that are used in the context of rituals and ceremonies associated with the life processes of community. They include theatre, dance-dramas, fairs, fiestas, and carnivals, oral narratives and other means by which collective memories are reinforced and tradition and community upheld. The informal includes all other forms of communication, for instance the use of proverbs, riddles, poetry, songs, etc., verbal and non-verbal, visual and auditory, used at times other than for ritual or ceremonial purposes.

However, the line between the formal and the informal in traditional communication may be drawn for purely descriptive rather than heuristic purposes, for it does not account for the linkages between the formal and informal media of communication. An example of this blurring is contained in the following richly evocative poem written by Basavanna, a leader of a major Hindu sect, the Virasaivites, quoted in Turner (1974: 281). It is simultaneously a religious poem as well as an instructional one. Through making use of the dialectic of structure and anti-structure, the poet affirms the relationship between the non-material, the self and the transcendental:

> *The rich*
> *will make temples for Siva.*
> *What shall I,*
> *a poor man*
> *do?*
> *My legs are pillars,*
> *the body the Shrine,*
> *the head a cupola*
> *of gold.*
> *Listen, O Lord of the meeting rivers,*
> *things standing shall fall*
> *but the moving ever shall stay.*[1]

Modernity's self-understanding of tradition

Much has been written and said about the crisis of the traditional. Accelerated processes of modernization along with the globalization of a unilinear model of development, of science and technology and of worldview and ethos, have affected the capacity of traditional cultures to reinvent themselves in the context of change. The modernizing nation-state has contributed to this process by homogenizing tradition, often along the lines of majoritarian identities. As a result, the traditions of minority communities, of 'first peoples' and of many marginalized groups are often bypassed in the making of a national tradition. This, in turn, has led to, on the one hand, the destruction of traditions or, on the other, has resulted in determined efforts by communities to assert their own traditions often through violent means of protest. The terms fundamentalism and revivalism are sometimes used to describe the actions of communities who resist modernity, secular ideals and a selectively defined global order. But these terms are used in a rather loose manner and they rarely shed light on the complex reasons that lie behind what are seemingly intractable, exclusivist positions.

Modernity has a history of rapprochement with tradition, although its motives and reasons have not been that transparent. The science of the 'Other' is contained within impressive, established canons of discourse. Numerous well-meaning Western sociologists and anthropologists of both mainstream and critical persuasions, have attempted to figure out the Orient, particularly its relationship with modernity and development. While this has certainly resulted in a wealth of micro and macro studies on every conceivable aspect of life in traditional societies, the boundaries of discourse have inevitably been coloured by a certain way of distinguishing 'them' from 'us'. Niranjana, *et al.*(1993: 1) in the introductory chapter to the volume *Interrogating Modernity: Culture and Colonialism in India,* have remarked on the politics underlying this process of 'othering': 'The mobilization of "Indian Culture" was as crucial to the West's construction of its identity in contrast to the Oriental Other as it was to the reconstructed Orient's attempts to define itself.' A sentiment re-emphasized in a more specific observation of O'Hanlon's (1989: 95) on the treatment of colonial histories of Hindu India: 'The underlying assumptions . . . of failure, in the proper separation between state and civil society, and of static tradition counterposed to the dynamism and maturity of Western political systems, have largely determined the framework within which the Indian past has been questioned.'

The question of representation has been a key element in the cultural-politics of a host of post-colonial thinkers including Edward Said (1978) and Homi Bhabha (1994), of critical traditionalists such as Ashis Nandy (1992) and Bhiku Parekh (1989), and critical modernists such as Jesús Martín Barbero (1993) and Néstor García Canclini (1993). While there is good reason not to take too seriously some of the more rabid, truculent and often ultra-exclusivist positions adopted by certain critics of modernity,[2] it is nevertheless important to reassess some of the assumptions reflected in the understanding of critical thinkers in the West. Take, for example, the German philosopher Jürgen Habermas, a critical modernist and defender (rightly so) of emancipatory traditions associated with modernity, its precious heritage of rights and freedoms and its potential for peace and justice. He has, on occasion, affirmed the need for a 'fusion of horizons' if the project of rationality is to succeed. And yet, in his classic study *The Theory of Communicative Action* Vol. 1, his defence of modernity is underwritten by an inability to empathize with or understand the logic animating what he calls the 'mythical worldview' (1981: 48). In Habermas's words: 'What irritates us members of a modern life world is that in a mythically interpreted world, we cannot, or cannot with sufficient precision, make certain differentiations that are fundamental to our understanding of the world. From Durkheim to Lévi-Strauss, anthropologists have repeatedly pointed out the peculiar *confusion between nature and culture'* (author's emphasis).

Such a reading militates against any fusion of horizons. What we are left with is yet another understanding of 'universals' that is built on a narrow conceptual base and on an inadequate dialogue with the varied particulars from around the world. These particulars are often animated by convictions based in a 'unity of feeling' rather than a 'unity of logic' (Fernandez, 1986), rooted in the symbolic, expressed through languages of relatedness that grasp the whole, and through a creativity that is forged in the dialectic of both 'structure' and 'communitas' (Turner, 1974). Gandhi, who believed in a similar self-understanding, once said that the act of liberation consists of both purifying the self as an aspect of serving the world and serving the world as a part of purifying the self, thus privileging relatedness. Johann Arnason (1991: 182) in a critique of Habermas calls our attention to what he terms the 'built-in conceptual obstacles that have to do with Habermas's vision of the distinction between the modern and premodern forms of thought'. For if, 'the specific characteristics of mythical and traditional worldviews are reduced to symptoms of an inability to grasp the differences between object domains (particularly those of the natural, the social and the subjective world) as well as between the ways of relating to them, there is no scope for an authentic fusion of horizons.' What is at stake here is a universal ideal of progress, inclusive of both its liberal as well as liberating variants.

There is a need to emphasize the reverse, to affirm authenticity in the particular and to come to an understanding of the universal based on a dialogue with the particular, especially with traditions whose self-understandings have evolved out of what one could term a 'relational dynamic'. In the *Bumba-meu-Boi*, a dramatic dance from Brazil, to the cult of the *Candomble*, also from Brazil, as well as the highly charged communicative experience of initiation during the rites of passage of the Ndembu from Zambia, to a whole host of traditions of communication from around the world, it is relational dynamics that enable meaning and the experiencing of it. The point that I have tried to make has much less to do with the particular self-understanding of a well-known social philosopher, i.e. Habermas, than with the fact that he is a representative of a tradition whose understanding of the 'universal' stems from a myopic and narrow reading and the denial of a universe of 'difference'.

The antinomies of traditional communication

The affirmation of relatedness is a central function of traditional communication. However, in itself this is by no means a virtue since relatedness may be the means of legitimizing a given order. The use of traditional communication in India, for instance, leads to the reinforcement of the status quo. While conceding the fact that in the present climate characterized by an accelerated assault on anything remotely believed to be 'traditional', justification of the status quo needs to be evaluated in the context of threats to the very survival of traditions, it would nevertheless be importunate for the 'progressive' in the traditional to remain a yardstick, a measure of its universality. Folk theatre, in its various manifestations in India, typically affirms through content and form the system, the continuities between Sanskritic Hinduism and 'folk' Hinduism, metaphyscial and material order and the functional inter-dependence of a hierarchial, caste-based social order. Through maintaining a system of patronage (Arden, 1971; Schechner and Hess, 1977), subsidizing entire castes whose primary function is to serve as village 'performers' (see Jayakar, 1980) and privileging the role of Brahmin priests in key rituals during pre- and post-performance, as well as through other means, the given order is preserved.[3] Wayne Ashley in a study of the ritual theatre of *Teyyam Kettu* from Northern Kerala, India, has commented on the many ways in which the *teyyam* performance affirms order:

> The performance serves . . . to perpetuate and renew a consistent social structure and distribution of power. This has been demonstrated by several examples: the transference of *Sakti* (power) and the consecration of the shrine by the Brahmin; the obligatory visit of the *teyyam* (consecrated idol) to the temple that houses the god of the higher caste; the allocation of special titles and gifts to the *kolakarran* (performers) from the king; the presence and the role of the Nair (high-caste) landowners; the calling out of the castes in order of ritual importance and the spatial delineations of the event. (Ashley, 1979: 111)

And yet, simultaneously, *Teyyam* and other folk media affirm the many bases for community. While the role of power and ideology may justifiably be highlighted, it is important to assess performances in the totality of their context. Traditional communication also affirms certainity, prevents anomie, establishes the basis for a normative order, legitimizes the role of the individual in society, strengthens solidarity across and between the various castes and other identity formations and reinforces traditions associated with collective memories. Clifford Geertz's observations on the role of sacred symbols in traditional societies may equally be used to describe the primary motive of traditional communication which is to 'relate an ontology and a cosmology to an aesthetics and morality: their peculiar power comes from their presumed ability to identify fact with value at the most fundamental level, to give to what is otherwise merely actual, a comprehensive normative import' (Geertz, 1973: 127).

Towards a critical reading of the traditional
It is often thought that the traditional implies something that is unchanging, fixed, static. Nothing could be further from the truth. The vitality of traditional forms of communication lies in their ability to negotiate with the current of change and to make appropriate modifications in their forms, narrative structures, characterization and content. For instance, electricity has been used to enhance both the stage setting as well as well as to heighten the appearance of characters in folk theatre and dance-dramas. Canclini has written about artisans from Ocumicho, Mexico who, faced with the alienating structures of modernity, reproduced ceramics representing devils, a means by which they came to terms with the symbols of modernity within their symbolic frames of reference. There are also innumerable instances from around the world of the adaption of content to context. During the struggles for independence waged in many parts of the South, peasant uprisings, the mobilization of the marginalized, and in a variety of 'social awareness' campaigns, traditional content was displaced in favour of 'critical' content, often with great effect. Rowe and Schelling (1991), Canclini (1993) and Barbero (1993), in their studies of popular culture in Latin America, have noted the trend towards the hybridization of cultures and the increasing reality of *mestizaje* (mixed) identities of urban migrants living in the cities of Latin America. These communities have had to

reinvent traditional crafts and cultures in order to preserve a measure of certainty in a volatile and fast-changing context. Taken in this sense, traditional forms of communication need to be seen as open texts that allow for both continuity as well as change.

Unfortunately, of late, there has been a trend towards the valorizing of the traditional, particularly its more exclusivist, discriminatory aspects. Often, majority communities have been responsible for consciously cultivating exclusivist understandings of tradition and of linking these to concepts of nationhood and identity. Thus the Serbs have assiduously used traditional communication to assert their identity as a Slavic people, as a nation and as the basis for citizenship. So have the Hindu revivalist groups in India for whom Indian tradition is nothing less than Hindu tradition. To an extent, this trend has gained some legitimacy from certain developments in neo-critical discourses, post-structuralism, postmodernism and post-colonialism that have willy-nilly supported the relativization of truth, objectivity and meaning. This has led conservative traditionalists to maintain that an evaluation of tradition may be conducted only from within the parameters generated by 'internal value systems'. The cultural critic Sarah Joseph calls our attention to some of the drawbacks of this mode of reasoning:

> An approach to tradition which relies solely on internally generated standards of rationality can be questioned on a number of grounds such as how we could recognise violations of internal standards, how we could understand the causal factors which give rise to those standards of rationality, and the possible conflicts and contradictions between different parts of social life. (Joseph, 1991: 60–1)

But it would only be right to point out that, often, such rigid approaches are a result and consequence of history. To communities who have been the victims of modernization, the retreat to certainty in tradition is the only means of upholding community at times of great political, economic and social upheaval.

Traditional communication and community

The strength of traditional communication lies in its capacity to visualize/pictorialize the expression of a vision of life, bridge the gap between its deep and surface realities and interpret this relationship in the context and idiom of the local. This is the means by which a particular vision of life is actualized and normalized. This vision is based on pragmatic, often non-dualistic understandings of life, as expressed in the following view taken by Ashis Nandy that it is often the case that in Third World civilizations, 'the concept of evil can never be clearly defined, that there is always a continuity between the aggressor and his victim, and that liberation from oppression is not merely the freedom from an oppressive agency outside, but also ultimately a liberation from a part of one's own self' (Nandy, 1978: 171).

This sense of pragmatism is deeply embedded in traditional forms of communication. Take for example the role of two folk theatre forms from India – the *Tamasha* from Maharashtra and the *Terrakoothu* from Tamil Nadu. Both play a role in the formation of consensus, the integration and reintegration of village traditions within the larger tradition of Hinduism; both, through their characteristic blend of satire, farce and ribaldry, make the connection between the sacred and the profane, the secular and the religious. Both provide the basis for the preservation of popular memory on a collective basis; both allow for an identification between the past and the present; both enable the Gods to be humanized; in both cases, their closeness to the lived life of people is their greatest strength. The reinforcement of caste divisions and power relationships is also part of their overall function.

In other words, the knowledge system which provides a framework for a *Tamasha* or *Terrakoothu* performance is both extensive as well as intensive. It is extensive in that it touches aspects of lived reality that are common, and that are manifested in daily life, i.e. problems related to love and anger, jealousy, etc. and their solutions which are framed within a construct of day-to-day morality. Furthermore, it teaches the community to be responsive to the need for generosity, tolerance and mutual respect. It is intensive in that it supplies a more or less complete means of identification with a system that provides security for both the individual as well as the community. It is a vision that is eloquently expressed in the cosmological drama of life, the dance of life personified in the image of the Hindu god Siva, as creator, preserver and destroyer of life.

Pragmatism is that which gives traditional visions their particular power, a pragmatism that is often achieved through the negation of the reality that they represent. In this sense, traditional forms of communication are also a means by which life is seen in perspective, a life that is conditioned by certainty in the transcendent as well as by

an indeterminate 'chaos' or 'fate'. This dialectic is alluded to by
Clifford Geertz in his description and comparision of the role of the
clown in the Javanese puppet theatre tradition of Wayang with that of
Falstaff in Shakespeare's Henry IV:

> Both figures . . . provide a reminder that, despite overproud assertions to
> the contrary by religious fanatics and moral absolutists, no completely
> adequate and comprehensive human world view is possible, and behind all
> the pretence to absolute and ultimate knowledge, the sense for the
> irrationality of human life, for the fact that it is unlimitable, remains.
>
> (Geertz, 1973: 140)

The source of strength of traditional forms of communication
often stems from their use of extra-linguistic modes of communication that privilege the imaginative over and above formally organized
grammars of communication. Such modes help illumine the many
ways in which society is built on mutually reciprocated relationships.
The mytho-poetic heritage of traditional societies is steeped in the
language of the imagination. Traditional forms of communication
enable the reinforcement of historical memories and consequently are
an eloquent testimony to the norm that the universal takes shape in its
encounter with the many different particulars. Moral and ethical
dilemmas, questions related to equity and justice and the distribution
and use of power are as much of concern to traditional societies as
they are to so-called modern societies. Traditional forms of
communication may not actively be used to challenge the given order
or the status quo. But like the mass media that have been on occasion
employed to challenge the dominant order, traditional forms of
communication have been used with great effect from time
immemorial to challenge authority, particularly regimes of repression
(see Appavoo, 1986).

Traditional forms of communication are part of a larger process
related to the making and re-making of communities. They play a
vital role in the process of negotiation that is itself a core element in
the self-understanding and growth of traditional communities. This is
an on-going process, but one that has become increasingly complex in
the light of the politics of change. It is this complexity that traditional
forms of communication endeavour to decipher and to make
intellible. Nothing more and nothing less. Against the juggernaut of
modernity and its tendency to homogenize difference in the name of
progress, traditional forms of communication are a gentle reminder

that true cultural democracy is forged in the interplay of difference, however idiosyncratic that might seem.

Notes

¹ The strength of the poem lies in the dialectics of structure and anti-structure, as for instance, between making (rich) and being (poor), between standing (sthavara) and moving (jangama), God and human beings, subject and object, leading to the final obliteration of all distinctions in the confluence of the rivers – the state of Samadhi.

² There are a number of instances from around the world of absolutists who have taken an uncompromisingly narrow view of matters related to ethnicity, faith and traditions. Bhargava's (1990: 56) assertion that 'people who claim to possess rights (must) operate within a culture of rights' is an apt counter to claims of exclusivity.

³ There has been a perceptible change in this situation during the last two decades. Various dislocations, economic, political and social have affected the relations of production in traditional communications. However, some of these functions are still visible in the folk traditions in different parts of the subcontinent.

References

Appadurai, A. (1986). 'Theory in Anthropology: Centre and Periphery' (356–61), *Comparative Studies in Society and History* 28.

Appavoo, J. T. (1986). *Folk-Lore for Change*. Madurai: T.T.S. Publications.

Arden, J. (1971). 'The Chhau Dancers of Puralia' (64–75), *The Drama Review*, 15, 2.

Arnason, J. (1991). 'Modernity as Project and Field of Tension' (181–213) taken from A. Honneth and H. Joas (eds.), *Communicative Action*. Cambridge: Polity Press.

Ashley, W. (1979). 'Teyyam Kettu of Northern Kerala' (99–112), *The Drama Review* 23,2.

Barbero, J. M. (1993). *Communication, Culture and Hegemony: From Media to Mediations*. London/Newbury Park/New Delhi: Sage.

Bhabha, H. K. (1994). *The Location of Culture*. London/Newbury Park/New Delhi: Sage.

Bhargava, R. (1990). 'The Right to Culture' (50–57), *Social Scientist*, 18,10.

Canclini, N. G. (1993). *Transforming Modernity: Popular Culture in Mexico*. Austin, Texas: University of Texas Press.

Fernandez, J. W. (1986) 'The Argument of Images and the Experience of Returning to the Whole' (159–87), taken from V. W. Turner and E. M. Bruner (eds.), *The Anthropology of Experience*. Urbana & Chicago: University of Illinois Press.

Geertz, C. (1973). *The Interpretation of Cultures*. New York: Basic Books.

Habermas, J. (1981). *The Theory of Communicative Action, Vol.1: Reason and the Rationalisation of Society.* London: Heinemann.

Jayakar, P. (1980). *The Earthen Drum.* New Delhi: National Musuem.

Joseph, S. (1991). 'Culture and political analysis in India' (48–62), *Social Scientist*, 19,10–11.

Nandy, A. (1978). 'Oppression and human liberation: Towards a Third World utopia' (165–180), *Alternative*, 4,2.

Nandy, A. (1992). *Traditions, Tyranny, and Utopias.* Delhi: OUP.

Niranjana, T., P. Sudhir and V. Dhareshwar (eds.) (1993). *Interrogating Modernity: Culture and Colonialism in India.* Calcutta: Seagull.

O'Hanlon, R. (1989). 'Cultures of rule, communities of resistance: gender, discourse and tradition in recent South Asian historiographies' (94–114), *Social Analysis*, 25, Sept.

Parekh, B. (1989). *Colonialism, Tradition and Reform: An Analysis of Gandhi's Political Discourse.* London/Newbury Park/New Delhi: Sage.

Rowe, W. and V. Schelling (1991). *Memory and Modernity: Popular Culture in Latin America.* London and New York: Verso.

Said, E. (1978). *Orientalism.* New York: Penguin Books.

Schechner, H. and Linda Hess (1977). 'The Ramlila of Ramnagar' (51–82), *The Drama Review*, 21, 3.

Turner, V. (1974). *Dramas, Fields and Metaphors: Symbolic Action in Human Society.* Ithaca/London: Cornell University Press.

9

The cultural frontier: Repression, violence, and the liberating alternative

GEORGE GERBNER

The new frontier of the struggle for democracy is the cultural frontier. Of course, traditional forces of inequity and injustice have had their cultural supports. But the cultural arms of new systems of colonization are now centralized, conglomeratized and globalized. They manufacture most of the stories for most of the children and discharge them into the common cultural environment. The mainstream of the environment, television, pervades every home and affects us every day from cradle to grave.

This is a report from that cultural frontier. It is in the spirit of Michael Traber's life's work, alerting us to the dangers of an increasingly dehumanized and constrained system of cultural mass-production and marketing. That system has taken the process of socialization out of the home, the school, and the church. Cultural policy-making has drifted out of the community and even the nation state, and out of democratic reach. We need to understand that process in order to liberate it from the constraints that, in the name of efficient marketing, twist it out of human shape.

In this report we shall review evidence from the ongoing Cultural Indicators (CI) project studying television content and effects since 1967. The CI database is a unique resource. It has detailed and coded observations including over 39,000 characters and 3,000 programmes in 1994. It yields our analysis of casting and fate, violence and victimization on television. The Cultivation Analysis contributes evidence about the consequences of growing up with television. We conclude with the results of our studies of what drives television violence and what we can do about it.

The typical viewer of dramatic network programmes on US television (now exported to almost every country) sees an average of

353 characters in prime time and 139 characters in Saturday morning children's programmes. Unlike life, fiction goes behind the scenes and shows how things work out in the end. Casting and fate reveal powerful moral and practical lessons. They demonstrate who is valued and why, who is likely to succeed and how, and who can get away with what against whom. Rarely, if ever, does a person encounter as many social types and relationships as often and in as compelling and revealing ways as on television.

The moral and behavioural lessons embedded in that synthetic world hold out great promise but also pose great dangers. Their aggregate facts and figures, remote as they may seem to be from everyday viewing experience, reveal what large communities absorb in common over long periods of time. That is the television everyone watches but nobody sees. Our children grow up and learn, and we all live, in the context of that world. Its patterns are repeated and confirmed every day, many times a day. It is resistant to change unless we know its contours and understand its dynamics.

Casting and fate in prime time

The world of US prime-time network television presents a coherent social structure that changes little over time. Men outnumber women three to one. Women tend to be concentrated in the younger age groups and 'age faster' than men. While 16 per cent of males but 25 per cent of females are portrayed as young adults, by the time they reach 'settled' adulthood, the proportions are reversed: 72 per cent of men but only 58 per cent of women are portrayed as settled adults. Men of nearly any age play romantic roles; their partners are younger women.

Romance may be rampant on prime time, but marriage is not. Only 11 per cent of all characters and 20 per cent of major characters are married. Marriage is a more defining circumstance for women than it is for men. More than two-thirds of all men but less than half of all women characters appear in roles whose marital status is indeterminate. Despite their generally younger age, women are almost twice as likely to play the role of wife as men are to play the role of husband.

Predictably, the population of prime-time television drama is overwhelmingly (about nine out of ten) 'middle class'. 'Upper class' characters are three to four times as numerous as 'lower class' characters.

The US Census classifies more than 13 per cent of the population, nearly one-third of the children of New York City, and one-third of all African Americans, as living in poverty. The US Bureau of Labor Statistics reports about 7 per cent of white, 15 per cent of African Americans and 20 per cent of teenagers seeking work as unemployed. Many more are low-income wage-earners. Avid viewers but poor consumers, they are all but invisible on television. Clearly identifiable 'lower class' characters make up only 1.2 per cent of all characters in prime time and even less in Saturday morning children's programmes. 'Lower class' women, who hold most of the lower-paid jobs in real life, are even more out of the picture; their percentage drops to nearly half of the men's in prime time and to one-third of the men's in Saturday morning children's programmes.

Race and ethnicity of prime-time characters is as skewed as gender, age and class, except perhaps African Americans. Their percentage increased in a twenty-year period, to over 11 per cent of all and 9 per cent of major characters. However, the representation of Latino/Hispanic American characters remained little over 1 per cent, Asian-Americans about 1 per cent, and Native Americans ('Indians') 0.3 per cent of all characters, and even less as major characters.

Positively valued characters (heroes) outnumber negatively valued characters (villains). 'Upper class' and Latino/Hispanic male characters have the largest proportion of villains, about twice the general percentage. The same groups, and 'lower class', disabled, gay/lesbian, and mentally ill characters have the highest negative ratio of 'good' vs. 'bad 'characters.

Ageing depresses the relative valuation of female characters. Women not only age faster than men but are also seen as relatively more likely to be evil as they age.

Heroes are destined to win and villains to lose, at least in popular fiction. Beyond that, however, being characterized as very rich, ill, or otherwise disabled is most likely to accompany failure. Characters depicted as mentally ill fail almost twice as often as they succeed, the highest ratio of failure in any group. Gender comparisons show that being old, 'lower class', lesbian, Asian or mentally ill places a special burden of relative failure on women.

Saturday morning children's programmes
The world of Saturday morning children's programmes magnifies all anomalies of prime time. Minorities drop in representation in

Saturday morning children's programmes, especially in major and female parts. Characters of the parents' generation, especially married and mother figures, are few and relatively ill-fated. There are few, if any, Latino/Hispanic, Asian/Pacific, or Native American females as major characters in twenty years of Saturday morning children's programmes samples.

The moral lessons of Saturday morning children's programmes are also more sharply delineated than those of prime time. There are more villains, and characters pay a higher price for heroism in that they have a higher ratio of 'bad' for every 'good' character. Older women and African American women bear the brunt of the relative devaluation.

The failure rate also rises with age until it reaches one out of four at the age of most viewers' parents. The relative balance of success vs. failure penalizes the old, the ill and disabled, and the poor.

The 'gender gap' heightens the inequities. Being relatively rare and 'bad' to begin with, older women are most likely to be depicted as deranged and to fail by a larger margin than in prime time. This is where the witches come from.

Violence

Violence can be seen as a legitimate cultural expression, even necessary to convey valid lessons about human consequences. Individually crafted and historically inspired, sparingly and selectively used symbolic violence of powerful stories is capable of balancing tragic costs against deadly compulsions. There is murder in Shakespeare, mayhem in fairy tales, blood and gore in mythology, although Greek drama, often cited for its compelling pathos and cathartic effects, showed only the tragic consequences of violence on stage. 'Greek sensibilities', observes theatre historian Oscar G. Brockett (1979: 98), 'dictate that scenes of extreme violence take place offstage, although the results might be shown.'

Under the increasing pressures of global marketing, however, graphic imagery is produced for world-wide entertainment and sales. This 'happy violence' is swift, cool, thrilling, painless, effective, and always leads to a happy ending, designed not to upset but to deliver the audience to the commercial message in a receptive mood. In this formula-driven dramatic fare, the limited degrees of attempted justifications for violence have been swamped in a tide of violent overkill and expertly choreographed brutality.

The marketing strategies driving mass-produced violence affect the total tone and context of programming. Beyond considerations of both quantity and quality, and above all other features and justifications, violence is a social relationship in which naked power is exerted. People hurt or kill to resolve a conflict, to force (or deter) unwanted behaviour, to dominate, to terrorize. Symbolic violence is literally a 'show of force'. It demonstrates power and shows who can get away with what against whom and at what cost to themselves.

Prime time
Our studies have found that violence extends the inequities of casting and fate. More major than minor characters commit violence, but minor characters, with their larger share of minorities, pay a higher price in victimization for the violence they commit. Latino/Hispanic and Native American characters, and those portrayed as poor, are the most likely to be involved in violence and to become victims of violence. In terms of a violence/victim ratio, 'lower class' characters pay the highest price: two victims for every perpetrator of violence.

Women generally pay a higher price in victimization for their violent actions than men do, and the price rises as they age. Older men suffer 182 victims for 100 perpetrators; older women suffer 215.

Lethal violence further extends the pattern. Characters of colour, Latino/Hispanic Americans, 'lower class', disabled or ill characters, and older characters are at the greatest relative risk of being killed instead of killing. The age differential strikes older women especially hard: they encounter lethal violence only to get killed. In short, men kill; women (especially older women) get killed.

Instead of muting the mayhem and inequities of prime time, Saturday morning children's programming intensifies them. More than half of all (including minor) characters are involved in violence, twice as many as in prime time. Eight out of ten major characters are involved in violence, compared to 52.3 per cent in prime time. The rate of retribution is also higher. For every 100 violent acts in Saturday morning children's programmes, there are 139 victims; for major characters the ratio is 127. Comparable ratios for prime time are 122 and 108.

Not only is there generally more violence in Saturday morning children's programmes, but minorities are disproportionately and mostly negatively affected. Native Americans ('Indians') and Latino/Hispanic American characters are the most violence-prone, significantly more than in prime time. The pattern is extended to the violence/victim ratio. African American characters suffer 108 victims for every 100 perpetrators of violence in prime time but 205 on Saturday mornings (whites suffer 135). 'Lower class' characters encounter violence only to be victimized; they have no power to inflict it. Asian-Americans pay the highest price: 267 victims for every 100 perpetrators (compared to 118 in prime time).

Although Saturday morning children's programmes present escalation of the pattern of prime-time violence in almost every age category, older women are again the most affected. Nine out of ten commit violence, and they absorb as much punishment as they inflict. They are evil, they are violent, they are the losers. Witches must die.

What are the consequences?

Cultivation Analysis ascertains what it means to be born into and grow up in a television home. Using standard techniques of survey methodology, questions about reality, security, feelings of vulnerability, and so on, were posed to samples of children, adolescents, and adults. The patterns of responses of heavy vs. light viewers, holding other factors constant, reveal the 'lessons' of growing up with television.

The 'lessons' range from aggression to desensitization and a sense of vulnerability and dependence. Victimization on television and real world fear, even if contrary to facts, are highly related. Viewers who see members of their own group have a higher calculus of risk than those of other groups, develop a greater sense of apprehension, mistrust, and alienation.

Heavy viewers in most subgroups are much more likely to express feelings of gloom and alienation than the light viewers in the same groups, and these patterns remain stable in surveys over time. Many subgroup patterns show evidence of 'mainstreaming'. For example, light-viewing men are less likely to express feelings of gloom than light-viewing women, while about the same percentage of men and women who are heavy viewers have a high score on this index. In other words, heavy-viewing members of the genders are closer together than light viewers of the two groups. Similar patterns hold when the associations are controlled for education and income. In short, heavy viewers seem to be more homogeneous, and more likely to express gloom and alienation, than their light-viewing counterparts.

These patterns illustrate the interplay of television viewing with demographic and other factors. In most subgroups, those who watch more television tend to express a heightened sense of living in a mean world of danger, mistrust and alienation. This unequal sense of danger, vulnerability and mistrust, and the homogenization of outlooks are the deeper problems of violence-laden, market-driven television. These are not simple policy issues. They are structural problems that any programme of change has to confront.

What drives television violence?

The standard rationalization is that violence is pervasive in television programming because it is popular. The evidence challenges the notion that violence is 'what the public wants to see'.

Of course, popular stars, strong stories and intensive promotion can make any programme a relative success, at least temporarily. Also, if only a small portion of the television audience gets 'addicted' to television violence, that can make graphically violent movies, videos, and games a commercial success. In fact, escalation of the body count seems to be one way to get attention from those addicted to global mayhem.[1] But that does not necessarily make violence *per se* popular with the television audience. Results of our comparative study of Nielsen ratings suggest that factors other than what the audience wants need to be considered to understand what makes violent programming profitable.

Is it popularity?

The A. C. Nielsen Company provides survey-based estimates of television viewing used by most broadcasters to set the prices charged for advertising time and to calculate 'cost per thousand' (CPM). CPM is the cost of reaching 1,000 viewers – the standard for assessing the relative marketing efficiency of different media and programmes, and the key economic factor in programming.

Nielsen rating is the estimated size of the audience viewing a programme, expressed as a percentage of the total sample. Share is the percentage of households tuned into a programme out of all households viewing at that time.

Two methods were used to compare Nielsen ratings and shares of violent and non-violent programmes. The first comparison samples were drawn by scanning all 30-minute and hour-length titles in the Cultural Indicators data base for five years, covering the 1988–9 to 1992–3 seasons. Violent programmes were defined as those that contained at least 10 seconds of overt physical violence per hour. Non-violent programmes had none. After eliminating titles that aired more than once within the same season's sample (in order to avoid undue emphasis on such programmes), each sample ended up with 101 programmes.

The second comparison eliminated programmes that were only occasionally violent, i.e. programmes that did not have violence in each annual sample. That comparison tests the ratings of repeatedly and consistently violent, occasionally violent, and non-violent programmes.

Comparisons of Nielsen ratings
The first comparison tests the general viewership of the total violent and non-violent programme samples. It shows that the overall average rating of the non-violent sample is 13.9 and the rating of the violent sample is 11.2. The shares of the non-violent and violent samples are 22.5 and 18.92, respectively. Furthermore, the non-violent sample is more highly rated than the violent sample for each of the five seasons tested.

The second method tests if there is a further difference between the viewership of repeatedly and consistently violent vs. only occasionally violent programmes. Programmes with some episodes that were violent and others non-violent are in a 'mixed' category. The remaining two categories contain consistently violent and always non-violent programmes.

This most rigorous test further demonstrates the relative unpopularity of violent programming. Non-violent programmes rate 17.2, mixed programmes rate 12.9, and always violent programmes rate 11.8. The respective shares are 27.7, 21.8 and 19.7. The gap between the relatively high viewership of non-violent and lower viewership of violent programmes increases with the increase of violence in the programmes.

The more consistently violent the programmes are, the more they decline in ratings, share, and presumably earnings based on them. The question arises that, as CPM is the key formula for longevity, perhaps violent programmes are sufficiently cheaper to produce than non-violent programmes to offset the loss of ratings. Therefore, the next assumption investigated was that controlling costs rather than increasing ratings may be an economic driving force behind violent programming.

Cost, genre, importance

Data compiled from the trade papers *Variety* and *Channels* (now defunct) show that the cost-control assumption is false. The average cost of non-violent programmes is $702,000, of occasionally violent programmes is $801,000, and of consistently violent programmes is $1,208,000.

The paradox of the persistence of violent programming despite low ratings and high cost required further investigation. It is possible that the programmes' genre rather than the presence or absence of violence accounts for differences in viewership. Ratings vary also by time period, as audience flow depends on the time programmes are aired. Finally, whether violence is incidental, significant, or the main focus of the programme might also affect viewing.

However, none of these potentially confounding conditions changes the results. The ratings gap favours non-violent programmes both before and after 9 p.m. Situation comedies that have some violence receive lower ratings and shares than those that have none. Crime-action programmes (where most violence is concentrated) are consistently rated lower than sitcoms and others. Humorous non-violent programmes have consistently higher average ratings and shares than mixed or serious programmes, and these ratings and shares generally decline as violence enters the programmes. Finally, as the significance of violence increases, viewership decreases.

Backlash

The trade paper *Broadcasting & Cable* editorialized (20 September 1993, p. 66) that 'the most popular programming is hardly violent as anyone with a passing knowledge of Nielsen ratings will tell you.' The violence formula is, in fact, a reason for popular dismay, political pressure, international embarrassment, and general institutional stress. Of course, growing up with violence produces its addicts who then provide the core audience for even more graphic cable programmes, movies, video games, etc. It only takes a small proportion of viewers, perhaps the equivalent of one night's television audience, to make other violent media a commercial success. But there is no evidence that, other factors being equal, violence *per se* is giving most television viewers in any country 'what they want'. On the contrary, most people suffer the violence inflicted on them with diminishing tolerance. Organizations of creative workers in media, health professionals, law enforcement agencies, and virtually all other media-oriented professional and citizen groups have come out against television violence.

A March 1985 Harris survey showed that 78 per cent disapprove of violence they see on television. A Gallup poll of October 1990 found 79 per cent in favour of regulating objectionable content in television. A *Times-Mirror* national poll in 1993 showed that Americans who said they were 'personally bothered' by violence in entertainment shows jumped to 59 per cent from 44 per cent in 1983. Furthermore, 80 per cent said entertainment violence was 'harmful' to society, compared with 64 per cent in 1983, reported Diane Duston of the Associated Press in the *Philadelphia Inquirer* (24 March 1993, p. F5).

'No topic touches a nerve in American homes as does violence on television . . .' began the lead article of a highly publicized special issue of *TV Guide* on 22 August 1992. Soon after, ten senators signed a letter to television executives demanding voluntary controls on violence. The Television Violence Act, in force since 1990, offered a three-year limited exemption from the threat of anti-trust action if the industry responded. It expired without evoking significant policy change.

Attorney General Janet Reno and Health and Human Services Secretary Donna Shalala, along with Department of Education Secretary Richard W. Riley, convened in Washington, DC, a 'National Consultation on Violence'. Their report, released in July, 1993, broke new ground in pointing out that: 'The issue of media violence is really just the first phase of a major cultural debate about life in the 21st Century. What kind of people do we want our children to become? What kind of culture will best give them the environment they will need to grow up healthy and whole?' The group recommended that citizens 'take lessons from the environmental movement to form a "cultural environmental" movement.'

By the end of 1993, President Bill Clinton and most members of the cabinet spoke out on television violence. No speech reverberated more than that of the Attorney General. 'Top cop Janet Reno may have turned Congress's anti-TV violence bandwagon into a runaway freight train', exclaimed *Variety* (1 November 1993, p. 25). Nine bills were introduced in Congress to curb television violence. A year later, none had even advanced to the floor of either house.

Meanwhile, local broadcast licence holders complained about their loss of freedom to choose what they show and exercise some control over violent programming. The trade paper *Electronic Media* reported, on 2 August 1993, the results of its survey of 100 television station general managers across all regions and in all market sizes. Despite the law that makes the licence holder fully responsible for programming for the local community and 'in the public interest', three out of four said there is too much needless violence on television; 57 per cent would like to have 'more input on programme content decisions'.

Networks were imposing their own programming formulas on affiliates, in clear violation of the letter and intent of the law and FCC (Federal Communications Commission) regulations. Even the trade paper *Variety* observed (22 August 1994, p. 19) that 'tough language in recent contractual agreements . . . is raising questions of whether the webs are playing fast and loose with long standing FCC rules mandating that stations – and not the networks – have the ultimate say in programme schedules.' For example, when, in the most dramatic media merger of 1994, Fox Broadcasting – the network owned by Rupert Murdoch's Australia-based News Corporation, 'financed', according to *Variety* (6 June 1994, p.1), '99 percent by foreign coin', and airing the most violent action shows – acquired the 12-station New World Communications Group, its contract stipulated that 'no (Fox) programming will be deemed to be unsatisfactory, unsuitable, or contrary to the public interest . . . which the licensee believes to be more profitable or more attractive', and none may be preempted 'except to present locally produced non-entertainment . . . approved by Fox.'

In an industry quick to claim the protection of the First Amendment when the violence formula is attacked, no loud voice was raised to protest against violations of broadcast licencees' freedom to choose programming most suitable to their viewers and the public interest. (It remained to the New York chapter of the NAACP (National Association for the Advancement of Colored People) to charge Fox with 'flagrant violation' of the FCC rule limiting foreign ownership of a broadcast station or network to 25 per cent. The reason was the web's cancellation of *Roc*, the only issue-oriented comedy about a working-class African-American family.)

Many in the creative community, however, expressed great concern about the loss of freedom. The Hollywood Caucus of Producers, Writers and Directors said in a statement issued on the eve of the August 1993 'summit' conference: 'We stand today at a point in time when the country's dissatisfaction with the quality of television is at an all-time high, while our own feelings of helplessness and lack of power, in not only choosing material that seeks to enrich, but also in our ability to execute to the best of our ability, is at an all-time low.'

Industry conflict and Hollywood's dissatisfaction was also reflected in a *U.S. News and World Report* poll, reported by the Associated Press on 30 April 1994. The Hollywood survey was conducted for the magazine by the UCLA Center for Communication Policy and found that views on violence inside the entertainment industry are not much different from those of the general public. The survey found that 59 per cent of Hollywood workers polled saw entertainment violence as a serious problem, 87 per cent said media violence is at least a contributing factor to violence in America, 58 per cent said they themselves have avoided violent programmes, and 76 per cent said they have stopped or discouraged their children from watching such programmes.

Leaders of the television industry responded by declaring their intention to run disclaimers and 'parental advisories', and, a year later, by commissioning violence 'monitors' to report still another year later. Another effort at damage control was the 'Industry-Wide Leadership Conference on Violence in Television Programming' in Los Angeles on 2 August 1993. It was dubbed the 'Violence Summit' by the international media crowding into its hotel ballroom. This was the first time that the electronic media industries invited legislators, educators, researchers and representatives of citizens' groups to discuss a matter of programming policy. The conference was covered by all major networks, broadcast live by CNN and later aired in full by C-SPAN. It made no effort to reach consensus, adjourned without making any recommendations for change, and had no impact on overall programme policy.

Nevertheless, industry sources cited in the trade paper *Broadcasting & Cable* (25 October 1993, p. 6) complained that 'we're not getting any credit for what we've already done.' Others called for a counter-attack and unveiled some of the most violent movies, programmes, and cartoon series ever produced. 'Up to now' said 'a network source' quoted by *Broadcasting & Cable,* 'we have tried to be good guys . . . I think you'll see a change in how we react.' A one-day 'snapshot' study of programming, reported in *TV Guide* on 13 August 1993, showed a significant rise of violence in the news, in promotional announcements, and in cartoons.

The global marketing factor
What accounts for the perennially violent fare, a virtual policy

paralysis in the face of the ratings and cost paradox, turmoil in the media industries, and fierce public backlash? The answer challenges the two standard rationalizations: first, that violence is what people want, and, secondly, that it is an expression of creative freedom.

Broadcasting & Cable magazine wrote in its editorial of 20 September 1993 (p. 66) that 'action hours and movies have been the most popular exports for years . . .'. Bruce Gordon, President of Paramount International TV Group, explained in the same journal (15 June 1992, p. 19) that 'the international demand rarely changes . . . Action-adventure series and movies continue to be the genre in demand, primarily because those projects lose less in translation to other languages . . . Comedy series are never easy because in most of the world most of the comedies have to be dubbed and wind up losing their humour in the dubbing.'

The magazine of the broadcasting industry returned to the theme in its 25 August 1994 'Special Report', entitled 'Action Escalates for Syndicators'. It noted that '. . . the closest thing to a guaranteed hit overseas continues to be U.S. action-adventure shows' (p.27). The most dramatic new entry into the 'action market' in 1994–5 is the *Action Pack* series produced by MCA TV, employing lavish special effects used in *Jurassic Park* and *The Mask*, and, despite its relatively good ratings, expecting a domestic deficit to be made up on the world market. Some executives, like Keith Samples, President of Rysher, a major syndicator of action programmes, have earned their 'reputation for negotiating international co-production deals that allow projects to succeed financially with lower domestic ratings . . .' (p.34)

Global syndicators demand 'action' (the code word for violence) because it 'travels well around the world', said the producer of *Die Hard 2* (which killed 264 compared to 18 in *Die Hard 1*). 'Everyone understands an action movie. If I tell a joke, you may not get it but if a bullet goes through the window, we all know how to hit the floor, no matter the language.' (Cited by Ken Auletta in 'What Won't They Do', *The New Yorker*, 17 May 1993, pp. 45–6.)

'Syndicators are developing action shows with international play in mind and are triggering 20 to 22 initial hours', *Electronic Media* reported in its 8 March 1993 issue (p. 4), because foreign buyers are 'tired of . . . series ordered in dribs and drabs of six or eight episodes – in genres they don't find appealing.' 'Action series' reported *Variety* on 5 October 1992 (p. 21) 'sell particularly well if produced by the dozens.' In today's trigger-happy market-place, a 22-episode order is a creative (and financial) cushion for producers 'because the network standard of 13 or even 6 instalments' is too paltry 'for cable and foreign markets where the marketers' profits come from'.

The answer to the dilemma of violent television programming thus rests in a highly concentrated and globalized system of production and distribution. Governments and private operators import violent action series in large quantities at low unit cost. The local product is typically more popular but, for smaller markets, much more expensive to produce.

US-based media industries dominate more than half of the world's screens, and violence dominates US production for export. A pilot study of international data in the Cultural Indicators database provides some information about the scope of the international 'overkill'. A thematic analysis of a sample of 250 US programmes exported to ten countries, compared with 111 programmes shown only in the US at the same time, found that violence was the main theme of 40 per cent of home-shown and 49 per cent of exported programmes. Crime/action series comprised 17 per cent of home-shown and 46 per cent of exported programmes.

Economic trends compound the pressures. Expensive and risky production requires the pooling of large resources and even larger distribution capabilities. 'Studios are clipping productions and consolidating operations, closing off gateways for newcomers', notes the trade paper *Variety* on the front page of its 2 August 1993 issue. The number of major studios declines while their share of domestic and global markets rises. Channels multiply while investment in new talent drops, gateways close, and creative sources shrink.

Concentration brings streamlining of production, economies of scale, and emphasis on dramatic ingredients most suitable for aggressive international promotion. Cross-media conglomeration and 'synergy' means that ownership of a product in one medium can be used, reviewed, promoted, and marketed in other media 'in house'. 'It means less competition, fewer alternative voices, greater emphasis on formulas that saturate more markets.' Privatization of formerly public-service broadcasting around the world means a decline of subsidies for the arts, reduction of staffs, and the production and distribution of more of the type of product that can be purchased at the lowest cost on the world market.

Networks pay producers a 'licence fee' for one or two airings of their product. The few buyers that dominate the market can set the licence fee so low that most producers do not break even on the domestic market. Deficit financing is the rule, not the exception, in programme production. This system places a great burden on producers and distributors. They must find additional sources of income to compensate for lower ratings and higher average cost of violent programmes and to make a profit. That is a difficult task that often takes a long time and demands a long-range strategy.

The additional sources of income are syndication of programmes, home video sales, various forms of ancillary merchandising and franchising, and, most importantly, foreign sales. The dependence on foreign sales affects the nature of the product in crucial ways. It makes producers search for an ingredient in a marketing formula that requires no translation, is image-driven, 'speaks action' in any language, and can be inserted into the culture of almost any country. They find that ingredient in violence. (Graphic sex is second, but, ironically, that runs into many more inhibitions and restrictions around the world.)

Production companies emphasizing alternative approaches to human conflict, like Globalvision Inc., G-W Associates, and Future Wave, report that they have difficulty selling their product. Far from reflecting creative freedom, viewer preference, citizen demands, or crime statistics, the global marketing strategy driving the television violence overkill wastes talent, restricts freedom, chills originality and damages human rights and the public interest. Helping broadcasters loosen these constraints, and serve audiences with more diverse fare addressed to their own needs and interests, is a key aspect of the cultural environment approach.

The cultural environment approach
Channels multiply but communication technologies converge and media merge. With every merger, staffs shrink and creative opportunities diminish. Cross-media conglomeration reduces competition and denies entry to newcomers. The coming of cable and VCRs has not led to greater diversity of product or actual viewing (see e.g. Morgan and Shanahan, 1991b; Gerbner, 1993b; Gerbner *et al.*, 1993).

A study of 'The limits of selective viewing' (Sun, 1989) related frequent thematic categories to the incidence of violence and found that, on the whole, television presents a relatively small set of common themes, and violence pervades all of them. A major network viewer looking for a nature or family theme, for example, would find violence in seven or eight out of every ten programmes. The majority of viewers who watch more than three hours a day have little choice of thematic context or cast of character types, and virtually no chance of avoiding violence. Fewer sources fill more outlets more of the time with ever more standardized fare designed for global markets. Global marketing streamlines production, homogenizes content, and sweeps alternative perspectives from the mainstream. Media coalesce into a seamless, pervasive and inescapable cultural environment, with television its mainstream, presenting a world that is iniquitous, demeaning, and damaging to those born into and living in it.

Media anti-trust legislation and broadcast regulations for localism, public trusteeship of licence holders, fairness and equal time, and against multiple, foreign and cross-media ownership and trafficking in stations are ignored, or obsolete, or irrelevant. There is no historical precedent, constitutional provision, or legislative blueprint to confront the challenge of the new consolidated controls that really count – global conglomerate controls over the design, production, promotion and distribution of media content.

The Cultural Environment Movement (CEM) was launched in 1991 in response to this drift. CEM is an educational non-profit tax-exempt corporation organized in the US to address the need to reach out internationally to build a coalition of independent organizations committed to joint action in developing mechanisms of greater public participation in cultural decision-making. It provides the liberating alternative to repressive movements in the field. It works to gain the right of a child to be born into a cultural environment that is reasonable, free, fair, diverse, and non-damaging.

Notes

[1]   The first rampage of *Robocop* for law and order in 1987 killed thirty-two people. The 1990 *Robocop 2*, targeting a 12-year-old 'drug lord', among others, slaughters eighty-one. *Death Wish* claimed nine victims in 1974. In the 1988 version, the 'bleeding heart liberal' turned vigilante disposes of fifty-two. *Rambo: First Blood*, released in 1985, left behind sixty-two corpses. In the 1988 release 'Rambo III' visits Afghanistan killing 106. *Godfather I* produced twelve corpses, *Godfather II* put away eighteen and *Godfather III* killed fifty-three. The daredevil cop in the original *Die Hard* in 1988 saved the day with a modest eighteen dead. Two years later, *Die Hard 2* thwarts a plot to rescue 'the biggest drug dealer in the world',

coincidentally a Central American dictator to be tried in a US court, achieving a phenomenal body count of 264.

References

Bandura, A., D. Ross and D. Ross (1967). 'Transmission of aggression through imitation of aggressive models', *Journal of Abnormal and Social Psychology*, 63, 575–82.

Brockett, Oscar G. (1979). *The Theatre; An Introduction*. New York: Holt, Rinehart and Winston.

Bryant, J., R. A. Carveth and D. Brown (1981). 'Television viewing and anxiety: An experimental examination', *Journal of Communication*, 31(1), 106–19.

Drabman, R. S. and M. H. Thomas (1974). 'Does media violence increase children's toleration of real-life aggression?', *Developmental Psychology*, 10, 418–21.

Eleey, M. (1969). 'Variations in generalizability resulting from sampling characteristics of content data: A case study'. MA thesis, The Annenberg School of Communications, University of Pennsylvania, Philadelphia.

Ellis, G. T. and F. Sekura III (1972). 'The effect of aggressive cartoons on the behaviour of first grade children', *Journal of Psychology*, 81, 7–43.

Gerbner, G. (1969). 'Dimensions of violence in television drama', in R. K. Baker and S. J. Ball (eds.), *Violence in the Media* (pp. 311– 40). Staff Report to the National Commission on the Causes and Prevention of Violence. Washington, DC: Government Printing Office.

Gerbner, G. (1970). 'Cultural Indicators: The case of violence in television drama', *The Annals of the American Academy of Political and Social Science*, 388, 69–81.

Gerbner, G. (1972). 'Violence and television drama: Trends and symbolic functions', in G. A. Comstock and E. Rubinstein (eds.), *Television and Social Behaviour, Vol. 1, Content and Control*. Washington, DC: US Government Printing Office, 1972, pp. 28–187.

Gerbner, G. (1972). 'The Violence Profile: Some indicators of trends in and the symbolic structure of network television drama 1967–1971', in Surgeon General's Report by the Scientific Advisory Committee on Television and Social Behaviour, Appendix A. (Hearings before the Subcommittee on Communications of the Committee on Commerce, US Senate, Serial No. 92–52.) Washington, DC: US Government Printing Office, pp. 453–526.

Gerbner, G. (1988a). 'Violence and terror in the mass media', *Reports and Papers in Mass Communication*, No. 102. Paris: Unesco.

Gerbner, G. (1988b). 'Television's cultural mainstream: Which way does it run?', *Directions in Psychiatry*, 8(9). New York: Hatherleigh Co., Ltd.

Gerbner, G. (1993a). 'Violence in Cable Originated Television programmes', a report to the National Cable Television Association.

Gerbner, G. (1993b). '"Miracles" of communication technology: powerful audiences, diverse choices and other fairy tales', in Janet Wasko (ed.), *Illuminating the Blind Spots*. New York: Ablex.

Gerbner, G. (1993c). 'Women and Minorities on Television; A study in Casting and Fate', a report to the Screen Actors Guild and the American Federation of Television and Radio Artists.

Gerbner, G. and L. Gross (1976). 'Living with television: The violence profile', *Journal of Communication*, 26(2), 173–99.

Gerbner, G., L. Gross, M. Jackson-Beeck, S. Jeffries-Fox and N. Signorielli (1978). 'Cultural indicators: Violence profile no. 9', *Journal of Communication*, 28(3), 176–207.

Gerbner, G., L. Gross, M. Morgan and N. Signorielli (1980a). 'Ageing with television: Images on television and conceptions of social reality', *Journal of Communication*, 31(1), 37–47.

Gerbner, G., L. Gross, M. Morgan and N. Signorielli (1980b). 'The "mainstreaming" of America: Violence profile no. 11', *Journal of Communication*, 30(3), 10–29.

Gerbner, G., L. Gross, M. Morgan and N. Signorielli (1982). 'Charting the mainstream: Television's contributions to political orientations', *Journal of Communication*, 32(2), 100–27.

Gerbner, G., L. Gross, M. Morgan and N. Signorielli (1984). 'Political correlates of television viewing', *Public Opinion Quarterly*, 48(1), 283–300.

Gerbner, G., L. Gross, M. Morgan and N. Signorielli (1993). 'Growing up with television: The cultivation perspective', in Jennings Bryant and Dolf Zillmann (eds.), *Media Effects: Advances in Theory and Research*. Hillsdale, N.J.: Lawrence Erlbaum Assoc., Inc., 1993.

Gerbner, G., L. Gross, N. Signorielli, M. Morgan and M. Jackson-Beeck (1979). 'The demonstration of power: Violence profile no. 10', *Journal of Communication*, 29(3), 177–96.

Gerbner, G. and N. Signorielli (1990). 'Violence profile 1967 through 1988–89: Enduring patterns', The Annenberg School for Communication, University of Pennsylvania.

Gross, L. (1984). 'The cultivation of intolerance: Television, blacks, and gays', in G. Melischek, C. E. Rosengren, and J. Stappers (eds.), *Cultural Indicators: An International Symposium*. Vienna: Austrian Academy of Sciences.

Hawkins, R. P. and S. Pingree (1982). 'Television's influence on social reality', in D. Pearl, L. Bouthilet, and J. Lazar (eds.), *Television and Behaviour: Ten Years of Scientific Progress and Implications for the 80's*, Vol. 2, Technical Reports. Washington, DC: Government Printing Office.

Krippendorff, K. (1980). *Content Analysis*. Beverly Hills, CA: Sage Publications.

Lovas, O. I. (1961). 'Effects of exposure to symbolic aggression on aggressive behaviour', *Child Development*, 32, 37–44.

Molitor, F., and K. W. Hirsch (1994). 'The effect of media violence on children's toleration of real-life aggression: A replication and extension of the Drabman and Thomas experiments.' Paper presented at the International Conference on Violence in the Media at St John's University, New York.

Morgan, M. (1982). 'Television and adolescents' sex-role stereotypes: A longitudinal study', *Journal of Personality and Social Psychology*, 43(5), 947–55.

Morgan, M. (1983). 'Symbolic victimization and real-world fear', *Human Communication Research*, 9(2), 146–57.

Morgan, M. (1984). 'Heavy television viewing and perceived quality of life', *Journalism Quarterly*, 61(3), 499–504.

Morgan, M. and N. Rothschild (1983). 'Impact of the new television technology: Cable TV, peers, and sex-role cultivation in the electronic environment', *Youth and Society*, 15(1), 33–50.

Morgan, M. and J. Shanahan (1991a). 'Television and the cultivation of political attitudes in Argentina', *Journal of Communication*, 41(1), 88–103.

Morgan, M. and J. Shanahan (1991b). 'Do VCRs change the TV picture?: VCRs and the cultivation process', *American Behavioral Scientist*, 35(2), 122–35.

Morgan, M., J. Shanahan, and C. Harris (1990). 'VCRs and the effects of television: New diversity or more of the same?' In J. Dobrow (ed.), *Social and Cultural Aspects of VCR Use* (pp. 107–23). Hillsdale, NJ: Erlbaum.

Morgan, M. and N. Signorielli (1990). 'Cultivation analysis: Conceptualization and methodology', in N. Signorielli and M. Morgan (eds.), *Cultivation Analysis: New Directions in Media Effects Research* (pp. 13–33). Newbury Park, CA: Sage Publications.

Pingree, S. and R. P. Hawkins (1980). 'US programmes on Australian television: The cultivation effect', *Journal of Communication*, 31(1), 97–105.

Rosenberg, M. (1957). *Occupations and values*. Glencoe, IL: Free Press.

Signorielli, N. (1987). 'Drinking, sex, and violence on television: The cultural indicators perspective', *Journal of Drug Education*, 17(3).

Signorielli, N. (1989). 'Television and conceptions about sex roles: Maintaining conventionality and the status quo', *Sex Roles*, 21(5/6), 337–56.

Signorielli, N. (1990). 'Television's mean and dangerous world: A continuation of the cultural indicators perspective', in N. Signorielli and M. Morgan (eds.), *Cultivation Analysis: New Directions in Media Effects Research* (pp. 85–105). Newbury Park, CA: Sage Publications.

Signorielli, N. and G. Gerbner (1988). *Violence and Terror in the Mass Media: An Annotated Bibliography*. Westport, CT: Greenwood Press.

Signorielli, N., L. Gross and M. Morgan (1982). 'Violence in television programmes: Ten years later', in D. Pearl, L. Bouthilet and J. Lazar (eds.), *Television and Behaviour: Ten Years of Scientific Progress and*

Implications for the 80's, Vol. 2, Technical Reports (pp. 154–74). Washington, DC: Government Printing Office.

Srole, L. (1956). 'Social integration and certain corollaries: An exploratory study', *American Sociological Review*, 21, 709–12.

Sun, L. (1989). 'Limits of selective viewing: An analysis of "diversity" in dramatic programming.' Unpublished Master's Thesis, The Annenberg School for Communication, University of Pennsylvania, Philadelphia.

Weiman, G. (1984). 'Images of life in America: The impact of American TV in Israel', *International Journal of Intercultural Relations*, 8, 185–97.

10

Linguistic minorities and the media

NED THOMAS

Soon after its invention, radio became a state monopoly in most European countries and indeed in much of the world outside the United States of America. When television became possible, its development followed the same pattern. When commercial stations were introduced, they in turn were licensed and regulated by the state. Both radio and TV have been important parts of the institutional cement that held the state together, reaching directly from the capital into the home, and perhaps in the process weakening some of the intervening institutions of society in the locality and the region. The media have been crucial in time of war in bolstering morale within the state, and as a propaganda arm abroad; where there are civil wars, the rebels have logically attempted an early take-over of the radio and TV studios. In peacetime the media have been seen as crucial to the proper functioning of democracy.

Within this state context, the typical debates about radio and television have concerned democratic access and balance, the right of all views to be heard; the depiction of sex or violence in programmes, the setting or undermining of standards; the balance between commercial interests and public service broadcasting, and the whole question of the relation of listening and viewing to trends in society – do the media educate or blunt sensibility? Are they agents of consumerism?

This chapter cuts into the material from a different angle. What does access to the media mean for those territorial linguistic groups who have not achieved their own nation-states? I am drawing on the experience of my own group, Welsh-speakers within the British state, but also on the comparable situations of linguistic minorities in France, Spain and other countries of the European Union. A wider horizon would produce further comparisons and some contrasts, but to give the argument weight and cohesion I must limit myself to those minorities which I know reasonably well.

The limitation of the discussion here to *territorial* linguistic minorities has another explanation. Immigrant minorities, while they may in some degree want cultural support in the host country from broadcasts in their own languages, also aspire to some degree of assimilation, and their languages do not live or die by the treatment they get in the host country; there is always the home country. Territorial linguistic minorities, on the other hand, experience the erosion on their territory of a life that was normal in that language; they tend to regard media and an education system in their own languages not just as cultural support for individuals but as ways of re-establishing linguistic normality and at least partly assimilating incomers to the language of the historic community on that territory.

The rise of linguistic and political nationalisms
The unitary nation-states of western Europe grew up on the basis of a single official language. The linguistic reality of Britain, France, the Netherlands and Spain was rather different. Military conquest and political union, discrimination against the minority languages in the courts, even the introduction of compulsory universal education in the official languages at the end of the last century and the beginning of this, failed to complete the process of linguistic assimilation, and languages such as Frisian and Basque, Catalan, Welsh, Breton, Irish and Scottish Gaelic entered the twentieth century with considerable numbers of monoglots and larger numbers of people who were happier in those languages and disadvantaged in situations where they were required to use the official language.

In my own country individuals with little or poor English would experience these situations as discrimination, yet it was not discrimination of the same kind that was experienced by Jews or black people. Welsh people could and did avoid discrimination by relinquishing the Welsh language, wishing to get on in the wide world of the British Empire by so doing or feeling an induced shame at speaking their own language. Induced inferiority-feelings in respect of language are a common feature of the history of minority-language-groups, and are still very evident among, for example, ordinary Breton-speakers. It is no coincidence that Breton should also have today only the most minimal presence on television. It is relegated by the media to the domain of folklore and made to appear unfitted for the uses of modern life as these are reflected on the TV screen.

When radio came into existence in the unitary Western European states, it naturally meant radio in the official language of the state (further afield in federal Czechoslovakia or the USSR, more than one language was used from the start). By the inter-war years, however, the stirrings of new linguistic and political nationalisms could be felt, very often stung into existence by the accelerating erosion of the unofficial languages. Radio and later television were felt by these new movements to be the most invasive weapons in the state arsenal. Minority languages have often achieved a temporary equilibrium with the official languages through diglossia – Welsh, for example, was long accepted as the language of chapel and home, alongside English as the language of commerce and education. But here were media that spoke the other language on the hearth, thus penetrating the ultimate refuge.

Radio became a politicized issue in Wales in the 1930s and a Welsh Home Service (with Welsh-language and English-language programmes) was won as a concession after a concerted campaign. A Welsh TV channel was the object of an eventually successful campaign of protest by Cymdeithas yr Iaith Gymraeg (The Welsh Language Society) in the 1970s, a campaign in which scores of people were imprisoned for non-violent direct action.

In Spain, the repression of languages other than Castilian was more severe and the reversal more sudden and dramatic. Basque and Catalan were banned from most public uses until close to the end of the Franco era. When a new democratic constitution was adopted in the 1970s the autonomous governments in these areas of the Spanish state lost little time in establishing television channels in their own languages.

In recent times, the reduced costs of TV technology, deregulation in some countries, and the greater tolerance of pluralism in general, have produced a second wave of minority-language TV stations. Scottish Gaelic has made great strides in relation to the small number of speakers. In 1994 a Frisian TV channel came on air and funding has been voted in the Irish Parliament for an Irish-language TV channel (Irish is a minority language in its own sovereign country though technically with official status).

In France, where the administrative boundaries are generally drawn in such a way as to cut across cultural boundaries, broadcasting has remained centralized, and the unofficial languages have been given only very marginal concessions in the media. At the level of private local radio stations, however, some progress has been made.

Where trans-frontier minorities exist, as in Swedish-speaking Finland, German-speaking Belgium or Slovene-speaking Italy, the minorities have sometimes been treated with relative generosity. The state will usually prefer its minority-language citizens to listen and watch services in their own language emanating from the state territory than that they should tune to services coming across the border in their language. But Austrian Slovenes, divided by a high mountain wall from Slovenia, receive scarcely any time on the Austrian channels yet are not enabled to receive programmes from neighbouring Slovenia. However, with the decreasing cost of satellite transmission and the European Union's declared open-skies policy, we may expect such situations soon to disappear within Europe. It has already been agreed that the new Irish-language channel shall be available throughout Ireland, North and South.

The growth of European institutions has, at least in the area of language, culture and media, favoured the rights of minority languages. Successive reports have been adopted by the European Parliament calling on member-states to grant full access to the media to their minority-language groups; and the Council of Europe, which has a wider European remit, makes the same demands of the states which sign its Charter of Regional and Minority Languages. Resolutions and charters do not in the short term produce change on the ground, but they do confer moral legitimacy and help forward change in the medium-term. The very concept of Europe, moreover, implies multilingualism. This emerging political unit cannot be integrated on the basis of a single language in the way that was attempted by the old nation-states. If there is to be a European citizenship, then people must be allowed to bring their languages with them into that citizenship; and how can one justify bringing some languages into the fold and not others? Clearly it is not numbers but structures of power that decide who is a minority: there are more Catalan-speakers than Danish-speakers, more Welsh-speakers than Icelandic-speakers. Once we begin to look at the great range of languages within the European framework, the hierarchy imposed by the nation-states begins to break down and we are confronted with a mosaic.

So far the story has been told in terms of an advance by the minority languages into the field of media from which they had been initially excluded. Though this has brought undoubted gains of a democratic nature, the story is more complicated than that. Languages that were kept alive by voluntary institutions and the everyday conversation of the people, and were in that sense deeply democratic, in entering the world of electronic communication become mediated through structures of political and economic power that have local, nation-state and international dimensions. Then again, many of those who campaigned for Welsh television in the 1970s must have imagined that it would be a means of disseminating more widely the largely amateur Welsh culture that then existed. But, of course, the men and women behind the cameras did not stand outside Welsh culture, neutrally recording it. They were active participants in that culture, transforming it into an electronic culture; and television needed full-time actors who set professional standards, and this again was very different from the part-time, amateur but participatory culture which had previously been known as Welsh. So the process of entry into the media has its own ironies and complexities, and it is with those complexities that we must now engage.

Strengthening minority languages through the media industries

A language at a given time is kept in existence by a group of people speaking to each other in that shared set of terms; and clearly in modern conditions, a language that does not have access to the media is doomed, for the media are an extension of people speaking to each other. So much that is new in the way of knowledge of the world today arrives via the media, and if it does not arrive in your language, then your language gets left behind, even at the very basic level of vocabulary. Such a stereotypically Welsh phenomenon as Rugby was at one time reported only in English, and Welsh-speakers habitually used English terms when discussing that game. Commentaries in Welsh have reversed that situation. The Welsh media are now major creators and disseminators of neologisms in a wide range of areas. In this way languages keep abreast of the modern world.

Media in one's own language also internationalizes the consciousness of the group. Instead of being given a regional or provincial slot within the media beamed from the capital, an audience that receives a full news service in Catalan or Welsh will get a mix of local, nation-state and international news as will those who watch news services in Spanish or English. This normalizes the status of minority-language speakers, raises their self-esteem.

Where literacy in the minority-language is poor, because of the language's exclusion from the education system, or for other historical reasons, as is the case with Basque, the coming of radio and TV in the language allows people to contribute to the national discussion who might otherwise not be able to do so. I was myself a reasonably articulate commentator on Russian affairs for Welsh television at a time when I could not, because of the circumstances of my education, have written a tolerably correct article in the language.

The presence of minority languages on television also raises their status among speakers of majority languages. Television is less opaque than radio, and most programmes have an eavesdropping audience of people who may not understand, or fully understand, what they are watching, but follow the pictures or the singing and perceive that here is a world by turns as interesting, humane, various, trivial or banal as the world of television in their own language. There is also evidence that television in minority languages is a stimulus and aid to the learning of those languages. All this can be very important not just for the health of the minority language but for peaceful coexistence in mixed-language communities.

Here we touch on the image which majorities have of minority-language communities. Each situation has its own special history but very rarely do minorities escape a stereotyping that mixes romanticization and contempt. While it may be true that all national and linguistic groups both suffer and inflict stereotyping, the situation is more serious for minorities, for they usually lack the power to project truer and more complex images of their life not only to the majorities but sometimes to their own people who may be tuned in to outsiders' views of the minority's life. Traditionally, stereotypes from outside have been challenged by the alternative versions of life provided by great writers in minority languages. In modern conditions, the media are crucial.

At this point we need to distinguish between transmission and production. The great majority of the world's films at present are made in the English language and in the United States. It is perfectly possible (and sometimes it is the only economically viable choice) to establish a television channel in your own language by setting up facilities for covering local news, sport and current affairs, filling the gaps with films dubbed or subtitled from other languages. The non-state languages of Spain in the immediate post-Franco era imported material on a large scale which they then dubbed; the aim was to have *Dallas* dubbed into Catalan before it was available in Castilian on a competing channel. In the early days this course was forced upon the new TV authorities because they had to build from a non-existent television base to a full channel in a very short time, but it did raise the question of whether television in one's own language might not also become a means of undermining one's own culture.

Welsh was more fortunate. The build-up to a Welsh channel was in stages so that a media industry existed in a small way and could be expanded when a whole channel was allotted to Welsh. But we also benefited in a backhanded way from our proximity to English. While the rest of the world could dub from English into their languages, Welsh people might well see the English-language programmes first on one of the British channels; so from the start, Welsh television needed to be more indigenous, to create its own programmes, including films. Something like 3,000 jobs were created in the Welsh media, at all levels, and the truth was brought home to us that what we had demanded as a cultural right was also an economic asset. Linguistic communities who are prepared to accept their media in another language in effect export jobs, export talent, and further weaken their own communities. This is a truth that has become evident to other linguistic minorities who are now building their own media industries.

Television films are an international market, and S4C, the autonomous Welsh channel, has sold into over seventy countries and languages throughout the world. Our own images of ourselves are thus projected to other people, where previously Welsh life was represented in films made by outsiders such as *How Green was my Valley*. It is not a question of outsiders necessarily being hostile: they may sometimes be sympathetic, they may romanticize you, but they will rarely know or show the *internal* tensions that a film made in the insider's language will reveal. Peoples such as the Indians of the Amazon forests are powerless indeed that are always the *objects* of others' study and gaze. At least with one's own media one has a degree of countervailing power.

But, of course, there is also a price to pay for belonging to an international market. Whereas radio usually stays within its own language-area, film and video can be sold around the world with 'voice over', dubbing and subtitling, and in the search for other markets one may distort the product by the criteria of the home market. Co-productions or back-to-back versions in two languages, can easily be influenced by the perception of the larger foreign market, so the battle for authenticity is not finally won simply because one has one's own media.

Minorities need to keep a constant watch on the evolution of new technology which can offer special opportunities for them or involve investment which is outside their capacity. Satellite television has been dominated by powerful languages, but as costs come down this may change, and it may soon be possible for citizens of Europe, wherever they work, to keep in touch with their home culture. But digitalization is on the horizon, and the investment required here may well, in the short run, limit its use to a few superchannels in the main languages.

Another area of concern for minorities is that of the news agencies, a question which has also been much discussed in the Third World context. The media in smaller languages cannot maintain a wide net of foreign correspondents and usually use material collected by one of the international agencies or a world-wide broadcasting organization such as the BBC. These typically have correspondents in the nation-state capitals whose coverage of minorities in those countries is sporadic, often sensationalist, and usually coloured by the perceptions of the nation-state capital. So the Welsh viewer is unlikely through this network to obtain any information or views that he or she would not encounter on English television about other minorities; and the view of the Welsh received by the Basque or Galician would similarly be coloured by the Spanish view. This effect is reinforced by the background resources available to newsroom journalists on CD-ROM. Thus, errors of fact which originally occurred in newspapers, and from which minorities are particularly liable to suffer, are called up on screen a hundred times and fed into innumerable articles and talks. What might have been an ephemeral lapse acquires authority when incorporated in a knowledge database. A *Sunday Times* caption interprets the word *eta* between two others on a girl's necklace as a declaration of support for a terrorist organization, when in fact it is the word in Basque for *and*, joining the names of two lovers; this interpretation then becomes available to other journalists round the world and the perception of the Basques as having virtually no existence except in the domain of terrorism is reinforced again and again. So far, schemes for alternative news networks among minorities have had no success, though bilateral contacts between minority media take place where the resources are available.

One development that runs parallel with the growth of minority-language media in Europe is the strengthening of regional media in the majority languages, as more and more states concede greater autonomy to the regions. It often seems as if the linguistically distinctive regions blaze the trail and are then followed by others. Regional media networks, when put together, have considerable economic power (as the *red autonómica* in Spain has proved), and it may be through such networks that linguistic minorities that have their own media will have the best chance of communicating with each other and with majorities in the future.

When nation-states have been intolerant of their minority-language speakers, it has often been on the grounds that they might turn out to be a disaffected fifth column and lead to the break-up of the state. This view can sometimes be self-fulfilling, since the more reason the minority has to be disaffected, the more subversive it may be. But is there reason to suppose that the growth of minority-language media results in a multiplication of political nationalisms? In western Europe the evidence does not suggest this. It is often a cultural nationalist vanguard that lays claim to media in its own language; but once established, the media belong to all members of that language group, nationalists and centralists alike. No language communities are monolithic, though they may sometimes appear so from outside; they have their internal tensions and divisions, and therefore the need to work out their differences.

In earlier times minorities dreamed of radio and television services in the image of those that existed in the dominant languages – powerful single services that commanded a mass audience and that wielded great influence, which they hoped would be used for the rehabilitation of their languages. When the explosion of minority media came, things proved rather different. The coming of the new language channels coincided with a wider proliferation of satellite and cable services, so that linguistic pluralism became part of a greater pluralism. In these conditions no groups are wholly imprisoned by their own national media.

Today, linguistic minorities in western Europe are by definition bilingual and can watch television in more than one language. Soon, perhaps, one may be able to say the same of majority audiences. Having your own media today means having a linguistic and cultural home among the burgeoning electronic signals available. What we now need to ensure is that an apparent pluralism of language and transmission channels is not dominated by an economic imperialism of production and technology. These pressures are found throughout the world, including many places where the indigenous peoples have far fewer resources than the relatively rich linguistic minorities of Western Europe. It may nevertheless be that our struggles here will have some relevance to them.

References

Davies, Janet (ed.) (1993). *Mercator Media Guide* Vol. 1. Cardiff: University of Wales Press.
Mercator Media Forum No. 1 (1995). Cardiff: University of Wales Press.

11

Mass media and religious pluralism

STEWART M. HOOVER

Relations between religion and the media seem nearly universally to be problematic. What we think of as the 'secular' media have developed a particularly diffident approach to the practices and claims of religion. At the same time, religious institutions and adherents seem to be wholly unsatisfied with and even suspicious of the practices and claims of the press. This is so even in the most self-consciously religious societies world-wide.

A variety of explanations for this situation have presented themselves. Most are rooted to a greater or lesser extent in understandings of the actions of contemporary actors in the media industries. The most common such claim is that the way religion is treated by the media stems from either ignorance or contempt on the part of media practitioners (cf. Lichter, Rothman and Lichter, 1982; see also Hoover et al., 1994 for a complete discussion). A related notion is that the worlds of 'religion' and 'the media' are somehow fundamentally incommensurate (Fore, 1987).

And why should we be concerned? Because of the challenges faced by contemporary cultures in accommodating themselves to increasing religious diversity. Global patterns of migration, as well as processes of social and cultural change in the North and the South are bringing about a situation where a multiplicity of religious expressions are coming into closer and closer contact with one another. Whether we see this as an evolving 'global ethnoscape', a consequence of 'religious postmodernism', or as a more mundane process of 'globalization', we now face conditions which require greater social understanding and cultural sensitivity. At the same time, our primary means of articulating and understanding these issues, the media, seem ill-prepared to carry out the task.

With this said, we must understand that the conceptual and analytical tools we have applied to the religion–media nexus have simply been inadequate heretofore. Hoping to build a sturdier basis for understanding, my purpose here is to begin the process of reconceptualizing the question in a way that takes account of history and, more importantly, fundamental social and theoretical considerations.

A question of history

Any approach to the problem of religion and the media must begin with the fundamental historical condition of modernity. As Marx, Weber, and others have pointed out, in the most general terms modernity has been typified by a movement from *organic* to *instrumental* relations. Organic relations were typical of an idealized past of closely-knit communities, where family, work, schooling, and religion were all physically and culturally 'close'.

Instrumental relations are typical of the still-evolving present age, and involve a disengagement of the relations of the organic era, and a loosening of 'bonds of affinity' that once tied communities together. Relations are now instrumentally and pragmatically organized. Intention plays a larger role in all social and cultural endeavours. A lingering hunger for the lost affinities of the past exists based both in a cultural memory of those bonds and in natural human needs for connectedness and belonging.

Significantly, there is an evolution in the mode of *cultural communication* typical of these two eras. In the *organic* era, communication is thought to have been more 'natural'. Storytelling and oral narratives predominated. Public communication was the province of the Church and the State (in various forms). Communication involved relatively little technology. In the *instrumental* era, communication has itself become more instrumental and 'artificial'. The printing revolution resulted in a major shift of authority over public communication.

Today, an independent institution of publicity and publication – the media – predominates, and the church and the state must submit themselves to this 'media sphere'. At the same time, the media constantly strive for the 'naturalized' status of the organic era. Their ability thus to become *tacit* constitutes one of their measures of success or failure. The media wish very much to become and remain transparent. Our cultural conditioning leads us to accord them the naturalized status to which they aspire. The traditional 'mirror metaphor' as an explanation for the relationship of media to social reality is an example of this.

Two environments

It is simplest to think of the situation as being one of two separate environments created and supported by the media. First, there is what we might call a *cultural* environment. In this environment we are asked to passively engage in the symbols, rites and rituals of mediated visual culture. This environment is enveloping and pervasive. The other environment is what we might call a *discursive* environment. In this sense and in this case, the media enforce an active role of 'publication', of 'making public'. It is perhaps too simple to think of these as 'entertainment' and 'news'.

Much can be said about the practices and prerogatives of *cultural* communication and of the culture industries. The enveloping liminal rituals of the entertainment media have been widely noted for their religious, spiritual, theological and more mundane or functional attributes. The *discursive* environment is, at the same time, also quite determinative of the prerogatives of religion, both formal and informal. As the question of the construction of religion in the media involves 'to a great extent' the prospects of the religion story in news content, this context deserves more serious attention than it has traditionally been given.

Contexts of discursive communication

It can be said that there are at least three types or contexts of communication in the discursive environment. The first of these is *private* communication. In this case, the communication takes place within a bounded setting, where participants typically know each other and where shared ideas and values can be assumed to be present. The signal example of this is a conversation. In private communication, both the terms and the extents of the communication can be controlled. Secrets can be kept, and private arrangements and agreements can be kept private. Boundaries between 'insiders' and 'outsiders' can be maintained, and are important points of negotiation and meaning. Group solidarity functions in relation to this private mode of communication, but solidarity can be maintained with reference to less private communication as well.

A second type of communication in the discursive environment is what we might call *community* communication. The typical examples of this are stories, narratives or sermons. In community communication, there is a closed and bounded group, but it is more of a collectivity than we assume with private communication, and a degree of anonymity of participants might be possible. In addition, community communication involves more of an exhortative or dramaturgic function.

The third type of communication in the discursive environment is fully public communication. Examples of such public communication include lectures, orations and announcements. However, social or civic rituals or dramaturgies are also public communication as are news, propaganda, advertising and public relations. Admittedly, there is a link between what I have called community communication and this latter category. In fact, as we shall see presently, I am convinced that there is little reason to make a distinction between these two any more, and that a difference in scale may be the only real difference in an age where mediated publication is the norm.

Religion in the media age

Religious practice and religious institutions now face an entirely new set of conditions as a result of the triumph of the cultural industries' control over the public sphere and public discourse. The most fundamental of these conditions is that the terms of engagement and even existence in the contemporary era are terms dictated by the conditions of the mediated public sphere. Whereas in the former 'organic' era, community-level consciousness and action were possible and were granted a good deal of space and scope, in contemporary times, community is constrained by the demands to accommodate to either a wholly private realm of individual action or to a public realm of discourse. Religious movements and religious institutions, which once existed quite comfortably in that 'middle', are now forced in either of these directions.

This situation has been recently described rather eloquently in the case of the United States by Yale University Law Professor Stephen Carter. In a 1993 book *The Culture of Disbelief*, Carter describes the attenuation of religion to these two 'poles' – the private and the public:

> . . . we often ask our citizens to split their public and private selves, telling them in effect that it is fine to be religious in private, but there is something askew when those private beliefs become the basis for public action.
>
> (1993: 8)

Carter's particular point – that religion in America is given more scope in private than in public, does not obviate the point that in contemporary terms, a middle, community-based ground, is no longer available. This can further be seen to be a logical consequence of the process of rationalization described by Weber (1963).

> The consistent message of modern American society is that whenever the demands of one's religion conflict with what one has to do to get ahead, one is expected to ignore the religious demands and act . . . well . . . *rationally.*   (Carter, 1993: 13)

But the situation is more than merely one of a rationalized religious sphere accommodating itself to the demands of enlightenment rationalism and the industrial order. The fact is that the most significant development of the modern period – the transformation in communication – has over the four centuries since Gutenberg wrought social and institutional changes which have massively changed the ground on which religion stands in the modern order.

Printing's impact was somewhat more profound and complex than that of an innovation in the distribution of books and the spread of literacy. As Elizabeth Eisenstein (1978) has shown, printing substantially changed both the technique of communication and the social and cultural structures through which publication, public articulation and public discourse occur. Its impact was not in the spread of books so much as it was in the spread of printing presses and the power that the ownership of the press conferred on a new class – the first 'knowledge workers' – publishers. Printing enterprises, according to Eisenstein:

> served as a kind of institute or activity . . . which rivalled the older university, court and academy and which provided preachers and teachers with opportunities to pursue alternate careers.   (Eisenstein, 1978: 191)

And, she concludes,

> [the] point of departure . . . is not the invention of one device in one Mainz shop but the establishment of many print shops in many urban centres throughout Europe over the course of two decades or so. This entailed the appearance of a new occupational culture associated with the printing trades. New publicity techniques and new communication networks also appeared.   (1978: 194)

The consequences of this development have been far-reaching. The contemporary inheritors of the social and cultural legacy of printing

– the media industries – now define and condition access to the public sphere in a way that cannot be escaped by any social institution. Religion and the state at one time held important positions in the definition of the public sphere and controlled access to its discourse. After printing, that power shifted. Today all social institutions, including the church and the state, must submit themselves to the conditions of the mediated public realm in order to exist in the public sphere.

The attenuation of the 'middle ground' of community in the postmodern era, combined with the rise of the mediated public sphere, thus creates a situation where religion can no longer exist on the boundary between 'private' and 'public'. In fact, religion has become, *de facto*, a public institution. Certainly there is always the possibility for subaltern discourses, communities, and movements to exist outside of the public realm, but that is precisely the point. To exist in any way other than as a closely bounded community well off the public stage is to cede the public realm to those symbols, values and ideas which are consciously and intentionally projected there, through what one religious figure of the turn of the century called '. . . the show windows of modern publicity'.

The conditions of the mediated public sphere

What is the nature of the challenge for presence and profile faced by religion in the contemporary era? As I have said, it is at its most fundamental a challenge to have a voice, to have presence in a public arena that is dominated by commercial and political speech. The terms of access to that arena fall into four primary categories: demands for (1) presence; (2) consensus; (3) plausibility; and (4) credibility.

In order to exist on the contemporary 'public stage', religion must first have a *presence* there. Often, such presence can, by itself, lead to certain values and significances. The conventional notion that the media have the power to 'confer status' on individuals, ideas or groups merely by presenting them has some validity. Public relations experts have been known to observe that '. . . there is no such thing as bad publicity', meaning that the value of presence alone is almost more important than the way in which one is treated by the media.

We also know that it is possible to achieve media presence by means other than what we might think of as 'honest' ones. Important and wealthy people receive more than their share of publicity merely by virtue of their status and power. This also holds for the world of religion. A whole generation of religious figures – now known as 'televangelists' – arose to political and discursive prominence in the US in the 1970s and 80s based on nothing more fundamental than their ability to purchase air time on television.

The second demand of the mediated public sphere is what I call *consensus*. This merely means that symbols and messages that find their way into that setting must 'make sense' in some fundamental way. They need to articulate with the general conceptions, ideas and values that are found in that realm. They cannot be too novel or too contradictory. They must fit preconceptions and understandings of both audiences and of practitioners of the media sphere.

Third, symbols and messages must have *plausibility*. That is, they must be seen to have reasonable or plausible explanations of reality. They must speak to both conditions and solutions (in the case of political or persuasive messages) that are accepted as real and meaningful. This is where the role and function of the media sphere can be seen and said to have theological implications. It is, in fact, on the level of plausibility that religions rise and fall, according to Berger's (1969) classic theoretical statement of the social construction of religion.

The fact that plausibility is under negotiation in questions of access to the media sphere places 'religion' and 'media' on a rather direct collision course. As we will see presently, contemporary religious evolution has left religion with few natural or a priori claims to legitimacy. That the conditions of the media sphere both establish the terms and context of legitimation, *and* can serve to provide competing symbols and cosmologies, and this means that in its very plausibility, religion faces a challenge more fundamental than merely its right to publicity. It is possible, even likely, that the realm of the media has now come to be the primary realm of the articulation and structuration of meaning for its audiences. A good deal of contemporary media theory stems precisely from this point. As new generations emerge, generations less steeped in the traditions of elders and more steeped in the symbolic customs of the modern, visual media, we are facing a time when the terms and language of religion are no longer relevant to the points at which people are living their lives. Practices of everyday life no longer involve legitimacies and explanations rooted in history and tradition. They are now self-legitimating, and the implications for traditional, institutional religion are rather stark.

The final condition determining access to the public sphere is *credibility*. Simply put, this means that the claims, symbols or messages presented in the public sphere must be *believable*. They must be unique claims that can be validated as true or at the very least not self-interested. This is the whole basis of journalistic practices of sourcing. All persuasive communications, including advertising, propaganda, political communication, and some types of religious communication, face the challenge of credibility.

In the case of religious or political messages, credibility is a particularly complex value. The news bureaux of the former eastern-bloc countries suffered from a lack of credibility, both internally and externally. The external broadcast services of most developed countries also have to negotiate a status for themselves that is rooted in their perceived credibility. Likewise, religious messages often face a presumption on the part of audiences that their primary purpose is propagandistic, and thus are not as credible as are messages from 'purer' journalist sources.

This means that, in fact, not all publicity is equal. Some kinds are better than others. The status-conferral of appearing in the media can be enhanced if one is presented there as the result of choice or selection by an unbiased third party, such as a journalist. Credibility can be harmed if one 'buys' one's own way there, or if a person appears to have special pleadings at heart in being there.

These threshold criteria thus establish a situation where religion must compete with other voices for access to the public sphere, and must compete on terms outside its control. Some religions have desired to move outside this system, to establish their own alternative networks. As I have said, that is all well and good, but in any context where it matters that religious messages get out (in cases of prophecy, evangelism, or social advocacy, for example) these conditions must be respected.

Conditions of contemporary religion

Today's religious climate also has important implications for relations between religion and the media. The characteristics of religious evolution in the industrialized West and in the North are important because the media of those regions play a determinative role in the traditions and standards of global media.

In the case of the most self-consciously 'religious' countries of the West – the US and Canada – a characteristic trend has emerged since the 1960s: the decline in support for all religious institutions, but particularly for the former 'establishment' religions of Protestantism. As has been noted elsewhere, the notion of 'secularization' has tended to dominate press attitudes toward religion. The secularization hypothesis holds that with industrialization and progressive rationalization, religion will gradually fade from view (Hoover *et al.*, 1989).

The decline of religion in the West has actually taken on the appearance of secularization, and decreasing media interest in religion over the past few decades can, in part, be seen as a reaction to the erosion of importance of formerly dominant religious institutions. However, far more is going on in the world of religion. In fact, religion has not faded but appears to be very much a feature of people's lives, even in those settings where it appears to be on the decline. While it has been suggested by some that it is best to describe religion as having become radically 'privatized' (cf. Luckman, 1967), others take a different view. Robert Wuthnow (1988), for instance, has described the current religious climate as one of 'restructuring' rather than decline or privatization. This restructuring has seen a gradual movement of religious activity away from the formal institutions toward more informal and para-church settings.

Recent work on 'baby boom' religion by Roof (1993) has further found that levels of religious interest remain quite high in contemporary life. Today's religious expression, however, is typified by what Hammond (1992) calls 'personal autonomy' in matters of faith, where the individual has come to be the active agent in the establishment and legitimation of religious practice. While the term 'cafeteria Catholicism' has been applied for quite some time to the way contemporary Catholics resolve the church's teachings on personal morality with practices which are at odds with those teachings, we might just as well use the term 'cafeteria religiosity' to describe much of contemporary practice across the board.

The implication of these trends for our considerations here is that individuals are now the *active agents* in the process of *legitimation*, and individual autonomy seeks out alternatives to the de-legitimated institutions and centres. This has led to the evolution of a powerful 'para-church' religious-cultural context with pretensions to the space being vacated by the churches. Thus the consciousness industries have evolved an autonomy of their own, and can provide a wide range of symbols, feelings, services and media to this new non-formal religious 'market-place'. Another way of describing this is to say that 'the lines cross': as formerly legitimate institutions fall, and as individual autonomy rises, a horizontal market-place of competing symbols, claims and institutions forms to fill the vacuum left by the departure of the formal groups and the traditional practices.

Characteristics of the new religious-symbolic 'market-place'

There are some specific characteristics of the context of religious symbols and expressions which have resulted from these trends. First, it is *rationalist* in the sense that its mysticism is tied to utility. The 'new religions' may stress mystical or spiritual matters, but they do so in order to establish a claim of their relationship to the individual and the individual's (often therapeutic) needs.

Second, this new context is *media-centred*. For the reasons I have already discussed, contemporary religious expression now takes place in a public sphere of discourse, not within formal, historic institutions. It depends on modern media of communication including direct mail, cassettes, satellites, fax machines, cable television, pamphlets, desktop publishing, etc. The media industries and the media sphere can and do provide a wide range of opportunities for meaning construction and for religious loyalty.

The new religious context is *individualistic* in the sense that it speaks to and for atomized, autonomous individuals. Community values and collective identity are not as important.

The new context is *syncretic*. It involves a practice that post-modernist theory calls *pastiche*: the construction of meanings, loyalties, and affinities out of a range of – often contradictory – images and appeals. Some analyses of the televangelism of the 1980s, for example, identified such a tendency in that movement (Hoover, 1988).

The new context is *synthetic*. It does attempt to weave these images and symbols into a 'coherent narrative of itself'. That narrative, however, in keeping with the general postmodern rejection of totalism, is one that is legitimated and based in particular constructions at particular times and never expresses pretension to historical or temporal orthodoxies.

The new context is *accessible*. That is, it is open to claims and to claimants based almost solely on technical characteristics such as market economics or technical sophistication. There is no process of legitimation which precedes entry into the mediated public realm.

Legitimation, to the extent it occurs, does so afterwards, in the experiences and practices of audiences.

Finally, the new context is *flexible* and *relevant*. It is able to speak dynamically and efficiently to real felt needs and interests. Its capacities in this regard are rooted in its practices of marketing. There is demographic and interest 'targeting' in para-church programmes and projects such as televangelism, the World Vision and Focus on the Family organizations, as well as in the 'secular' media which are targeted at these same issues. Audiences are sought and located 'where they are'. Audiences and adherents are no longer required to go to a specific institutional setting to do their religious business. Churches and para-church organizations 'go to them' instead.

The best example of this latter phenomenon is the so-called 'megachurch' movement in North America, where very large quasi-evangelical congregations are emerging, which combine an orientation toward service of needs and interests with a 'soft-sell' evangelism and a media and marketing-oriented appeal. The largest of these congregations have memberships in the thousands and very directly eschew any linkage to traditional denominations or historical religious institutions.

Implications

I return now to the beginning. How are relations between religion and the media to be reformed in the years ahead? It is clearly the case that the construction and treatment of religion by the 'secular' media is inadequate for an era of increasing importance and diversity of religious expression. As we have seen, for a variety of reasons, both religion and the media are profoundly ill-prepared to address this situation. It is not enough though merely to 'blame the media'. It has been my purpose here to attempt to describe how the situation is one that has important social and historical roots, and that it is rooted in – not a direct cause of – more profound changes that are gripping religious institutions and religious practice in the postmodern media age.

Prescriptions for improved media understanding of religion have been proposed elsewhere (Hoover *et al.*, 1989; Hubbard, 1990; Dart and Allen, 1993; Hoover *et al.*, 1994). These, in sum, describe the ways in which religion in public discourse is a function of media practice, and suggest approaches to the problem that should better serve societies and the globe.

But what of religion? What can and should it do? If we are talking about what I have called the 'cultural' communication of the commercial, global, entertainment media (often called the 'culture industries') then the issue is rather striking. The engine of mediated communication is providing a powerful challenge to the traditional prerogatives of religion to legitimate and define understandings of the nature of reality, of faith and of spirituality.

In the area I have called 'discursive' communication, traditional religion can and should play more of a role (a role which could serve to address the culture industries as well).This role will be an historic reversal for churches which have traditionally ignored, denigrated, and trivialized the media sphere and its importance. If such a turn-about is possible, however, some fairly specific and pragmatic considerations present themselves.

At their most basic, these considerations revolve around the moment when religion chooses to enter the mediated public sphere and raise a set of questions which should enforce a clear and rational calculus to the process. Religion must decide if it is interested primarily in *private, community,* or *public* communication. The implication of each is clear. To privatize is to disappear from the discourse. The option of community-oriented communication is increasingly unavailable, as more and more adherents and potential adherents are embedded in the media sphere rather than in local community. The risks and challenges of public communication have already been discussed at length.

Religion must also decide *where* its 'audience' *is*. This enforces a set of considerations that are outside the normal purview of religious and theological logic, but which are rooted radically in the logic of the culture industries. Religion must, finally, decide how to get to those audiences and with what messages.

It has not been my purpose here to make a theological case for these various conditions and practices. In fact, it has been noted by many that in some ways the practices of the media age are directly contradictory to fundamental theological claims of the Abrahamic faiths. This is not a trivial question, obviously. However, reticence rooted in theology has, for too long, kept religious institutions from taking seriously the challenges of the media age. They must now consider whether, without understanding these challenges, they will have missed the opportunity to decide how and under what circumstances they survive in the public sphere in the next century.

Acknowledgement
Preparation of this chapter was supported in part by a grant from the Lilly Endowment, Indianapolis, USA.

References
Bellah, Robert, Richard Madsen, William Sullivan, Ann Swidler and Steven Tipton (1985). *Habits of the Heart: Individualism and Commitment in American Life.* Berkeley: University of California Press.
Berger, Peter (1967). *The Sacred Canopy.* Garden City, NY: Anchor.
Carpenter, Joel (1985). 'Tuning the Gospel: Fundamentalist radio broadcasting and the revival of mass evangelism, 1939–45.' Paper delivered to the Mid-America American Studies Association, University of Illinois, Urbana, 13 April 1985.
Carter, Stephen (1993). *The Culture of Disbelief.* New York: Basic Books.
Dart, John and Jimmy Allen (1993). *Bridging the Gap: Religion and the News Media.* Nashville, TN: Freedom Forum First Amendment Center.
Eisenstein, Elizabeth (1978). 'In the wake of the printing press', *Quarterly Journal of the Library of Congress* 35:3 (June), 183–97.
Fore, William (1987). *Television and Religion.* Minneapolis: Augsberg Press.
Hammond Phillip E. (1992). *Religion and Personal Autonomy: The Third Disestablishment in America.* Columbia, SC: University of South Carolina Press.
Hoover, Stewart, Shalini Venturelli and Douglas Wagner (1994). *Religion in Public Discourse: The Role of the Media.* (Mimeographed report). The University of Colorado.
Hoover, Stewart, Barbara Hanley and Martin Radelfinger (1989). *The RNS-Lilly Study of Religion Reporting and Readership in the Daily Press.* (Mimeographed Report). New York: Temple University and Religious News Service.
Hoover, Stewart (1988). *Mass Media Religion: The Social Sources of the Electronic Church.* London: Sage.
Hubbard, Ben (1990). 'The importance of the religion angle in reporting on current events', in B. Hubbard (ed.), *Reporting Religion: Facts and Faith.* Sonoma, CA: Polebridge Press.
Hunter, James Davidson (1991). *Culture Wars: The Struggle to Define America.* New York: Basic Books.
Lichter, Linda, Robert Lichter and Stanley Rothman (1982). 'The once and future journalists', *Washington Journalism Review,* December, pp. 26–7.
Luckman, Thomas (1967). *The Invisible Religion.* New York: Macmillan.
Marty, Martin and Scott Appleby (1992). *The Glory and the Power.* Boston: Beacon Press.
Mattingly, Terry (1993). 'Religion in the news: Are we short-changing readers and ourselves with biases that filter news?', *The Quill,* July/August, pp. 12–13.
Roof, Wade Clark (1993). *A Generation of Seekers.* New York: Harper Collins.

Schultze, Quentin (1987). 'The mythos of the electronic church', *Critical Studies in Mass Communication* (4), pp. 245–61.

Tocqueville, A. de (1945). *Democracy in America*. Edited by P. Bradley in 2 volumes. Vol. 1. New York: Alfred A. Knopf. (Original work published in 1834 and 1840.)

Weber, Max (1963). *The Sociology of Religion*. Translated by E. Fischoff. Boston: Beacon Press. (Original work published in 1922.)

Wuthnow, Robert (1988). *The Restructuring of American Religion*. Princeton: Princeton University Press.

12

Communication: international debate and community-based initiatives

CARLOS A. VALLE

Beginning with the rapid development of technology, its enormous impact on the social life of our planet, and the political, economic and social changes that have taken place in recent times, communication media have come to play a preponderant and indispensable role in modern life. Hence any consideration of their place in society has to be done within a wide-ranging spectrum that takes into account the nature of the media in the context of the political, economic, cultural and social structures in which they develop as well as the phenomena of globalization and localization that we are experiencing today.

In order to try to understand the current international debate on communication at the level of the community, it is important to examine one or two basic aspects of recent history.

A short and challenging history

'The global village', the celebrated term coined by Marshall McLuhan (1967), defined what he understood to be the modern phenomenon of communication on the basis of technological development and the proliferation of the mass media. It has been, and still is, frequently quoted, but is it right? Has the world really been recreated in the image of a global village by the influx of technology? Or in the image of what is actually changing our world?

In 1977, when the MacBride Commission, to which I shall refer later, began its work on the search for a new information and communication order, it had a considerable amount of information available to it. For example, the final report reveals that the increase in the number of radio and TV sets in the Third World in the period 1960–76 was enormous. In Africa there were seven times as many

radio sets and twenty times as many TV sets. In Asia the ratio went
up from one to five. In Latin America, there were four times as many
radio sets and almost ten times as many TV sets.

If we relate these statistics to the period 1950–75 and look at the
figures on a world scale, the following can be shown. The percentage
increase in the number of radio sets rose to 417 per cent, while the
number of TV sets increased by 3,235 per cent. This gives a clear idea
of the exponential development of the mass media.

Pausing for a moment, we might be tempted to say, along with
McLuhan, that the world really is being turned into a global village.
But things are not so simple when we consider the statistics and
compare regions. In 1976, while in North America there was one TV
set for every two people, and in Europe one between four, in Latin
America there was only one between twelve; in Asia and the Arab
countries one between forty; and in Africa just one between five
hundred.

Looking at this in a broader context, adding information regarding
the industrial development and concentration of power in the media
sphere, the following becomes apparent (Guía del Tercer Mundo,
1988). The Third World provides only 10 per cent of global industrial
production and in more than twenty years, its industrial output
increased by just 2.3 per cent, from 8 per cent to 10.3 per cent. But
that's not the end of the problem. Practically a quarter of that
amount is concentrated in just one country, Brazil.

If we consider that about 100 countries can be counted among
those belonging to the Third World, then 73.2 per cent of industrial
production is concentrated in just ten of them. In itself this situation
is a constraint on any kind of advanced technological development.
To this must be added the fact that these countries are exporting
primary materials and importing manufactured goods.

The problem with this kind of exchange is well known. The
countries of the Third World, by importing manufactured goods, have
to pay for the enormous cost of salaries in the industrialized countries
out of the resource of their low incomes. Thus, while the average
earning in underdeveloped countries is of the order of $800 a year, in
the countries of the North the figure rises to $10,000. But the problem
is still more serious. If the population of the underdeveloped
countries is estimated at some 2,300 million, it becomes obvious that
about 2,000 million people earn less than $500 dollars a year.

In search of a response

How has the world community reacted, faced with this reality? What has been proposed and what has actually been done?

The history of the response to these problems indicates possible solutions but also the enormous difficulties encountered. This history revolves around two ideas: the search for a New International Economic Order (NIEO) and the search for a New World Information and Communication Order (NWICO). Although I shall refer in particular to the latter, we have to remember that both share a history in common.

The NWICO debate began in the 1970s, on the initiative of the Third World countries, with the idea of reducing the gap between the rich and poor countries. It was based on the conviction that growth in itself is not a panacea for a solution to the countries' problems nor for sustainable development, and that the free market is not the most effective mechanism to ensure adequate distribution of resources. In one sense all of this began with the birth of the so-called 'Group of '77', which was formed on the occasion of the first UNCTAD conference in 1964. But as Grifeu reminds us: 'The global political strategy was designed and co-ordinated by the Non-Aligned Countries' Movement, founded at the Bandung Conference in 1955' (Grifeu, 1986: 82).[1] This group would become one of the great defenders of the need for a new international information order, considering it fundamental that any proposal for a new economic order must go hand in hand with an adequate new communication order.

Four basic elements in the approach to the search for a new economic order can be highlighted (Estévez, 1993).

- Denunciation of the intrinsically unjust nature of the international economic system and the need to change it, which carries with it the need to protect states and to control the multinationals.
- National appropriation of a greater part of income from the land, agricultural and mineral, likewise establishing the right of national ownership.
- A new impetus towards the industrialization of the Third World, by means of the transfer of financial and technological resources, controlling the activities of the transnationals and promoting greater South–South exchange.

- Strengthening and developing the sovereignty of Third World countries, guaranteeing the central role of the state.

These objectives underline the close relationship between the economy and modern communications and the urgent need to begin the search for a new information and communication order. It was in 1977 that an important step was taken when Unesco approved the creation of an International Commission for the Study of Communication Problems, which would become popularly known as the MacBride Commission in honour of its President, Seán MacBride, Irish statesman and winner of the Nobel Prize and Lenin Peace Prize.

The MacBride Commission presented its final report, called *Many Voices, One World*, to the Unesco meeting in Belgrade in 1980.[2] As a report that intended to do justice to different positions, it was ambiguous, contradictory and deficient. However, its contributions laid the foundations for an in-depth discussion of the current situation and a beginning to the search for a more democratic future leading to greater independence and self-development of local cultures.

The implementation of the NWICO quickly came to be resisted. At the end of 1984, the United States, soon to be followed by Great Britain and Singapore, withdrew from Unesco, alleging among other things that the NWICO represented an attempt to restrict the freedom of the press and private enterprise.

According to Reyes Matra (Grifeu, 1986: 169) the offensive launched by the US against the NWICO, before its withdrawal, concentrated on three fronts: discrediting the endeavours of Unesco and the Third World and returning the debate to technological needs and technical assistance; discrediting multilateral agreements and emphasizing bilateral agreements; bringing pressure to bear on governments and the private sector in a return to the old strategy of the 'free flow of information'.

Given that the United States was supporting a considerable portion of Unesco's general programme, its withdrawal became a kind of economic sanction which would oblige those who endorsed NWICO to rethink the inconvenience of continuing to do so and to think seriously about any similar proposal in the future.

From that date onwards the destiny of the NWICO, at least at the heart of Unesco, began to seem futile. The flags in favour of a New Order were gradually furled and replaced with new ideas. This became evident at the 24th General Conference in 1987 when Federico Mayor, of Spain, was elected as Director General of the organization. He began by adopting a particular stance with regard to international communication, stressing that Unesco 'is committed to defending the free flow of information' and, later, insisting that it rejected the concept of a 'new international information order' (Traber and Nordenstreng, 1992: 6). During the 25th General Conference, despite the calls of various developing countries, the Plan for the years 1990–5, which outlined the activities for that period, only mentioned the NWICO in its introduction and not in the operational sections. Unesco's 'new strategy' concentrated on emphasizing the free flow of information and the freedom and independence of the media; priority was given to operational activities as against research, and considerable weight was given to the role played by information technology in development. In the preparatory documentation for the 27th General Conference (October 1993), among other things it was proposed 'to approve the emphasis given to encouraging press freedom and the independence and pluralism of the media in all regions' (Unesco, 1993).

Unesco's change of direction was a forewarning of radical changes in its programmes. For it was expected that these new proposals would open the door to a return of those who had previously abandoned ship. Up to now this has not happened, and there is no indication that in the short term it will.[3]

Approval of the Third Medium-Term Plan 1990–5, whose Programme IV focused on 'Communication in the Service of Humanity', indicated that the call for a new order was not a 'spur-of-the-moment invention' and that it was necessary to 'take the lessons of past experience to heart and to explore the possibility of a new strategy'. Moreover, the document insisted on a specific objective: 'to ensure free flow of information at international as well as national levels, and its wider and better balanced dissemination, without any obstacle to freedom of expression, and to strengthen communication capacities in the developing countries, so that they may participate more actively in the communication process.' In order to carry out the programme, work would concentrate on training. It seems that there was no room for 'the lessons of the past'.

Herbert Schiller wrote that, beginning with the programme mentioned above, the movement towards a new order in Unesco went into eclipse, while the issues themselves remain more urgent than ever

(Schiller, 1990).[4] For him, the discussion will take place with or without Unesco's support.

Where are we and where are we going?

The enormous changes recently experienced throughout the world have produced political and intellectual paralysis in vast areas and a kind of conformist position-taking. The fall of the Berlin Wall, symbol of the disintegration of the socialist bloc, marks acceptance of the end of an era and conviction that today's world is controlled exclusively by the market. 'The end of history' is a slogan that has won over some converts. They argue that there is no longer any tension between East and West. Communism has been overthrown. Capitalism has finally triumphed. From now on the market economy will prevail and determine the life and progress of all people. Therefore, they say, we must accept and live this reality, a reality not made up of dreams and promises of change, but of concrete events and possibilities within easy reach of those willing to adapt to the new situation.

We should recall how, at the beginning of the 1960s, with the influx of prospects of change in many parts of the world, three intellectual paradigms related to notions of development, cultural imperialism, and cultural pluralism dominated both the communications debate and the proposed solutions. Disillusion and frustration in many quarters put those paradigms in the shade. They were criticized from different points of view, especially for their political connotations and for having demonstrated their inability to produce results.[5] Such critiques often ended with wordy proposals about the importance of focusing on the 'problem' of culture or enquiring into giving new emphasis to 'civil society', but usually avoided serious reflection about the 'lessons of the past'. Many intellectuals seemed to take refuge in the search for ideas adapted to the prevailing context rather than the search for new visions that might gather up what was learnt in the past and face up to issues that had not been resolved. It's as if they were making a qualitative leap in order to throw a blanket over the past and try to draw up new paradigms that might avoid taking positions that inflamed the existing general panorama.

We can agree that the ideas set out in the search for a new order are not necessarily the best for the present day and that it is necessary to revise paradigms that tended to become immovable dogma in order to see present reality with fresh eyes. However, valid questions remain, such as whether the problems that gave rise to the initiative are still there, whether or not those problems were faced and resolved, how and why and what new claims are emerging today.

When the MacBride Commission began its work it had, as a basic proposition, to offer a response to five fundamental problems. By enumerating them we have to ask ourselves where we are today and in which direction are we moving.

Dependence of the non-industrialized on the industrialized

The reality is one of dependence by Third World countries on the industrialized countries in terms of technology and their increasing integration under the sway of multinationals whose basic interest is profit. This situation of growing dependence brings as a result a rapid deterioration in local cultures and the creation of values that reinforce dependency and submission to the dictates of the new culture of consumerism.

While agreeing that dependency theory does not respond adequately to this situation, we cannot conclude that it is no longer present in the world. The problem of communication and power is perhaps the most pressing of all. It is sufficient to quote Ben Bagdikian on this point:

> A handful of mammoth private organizations have begun to dominate the world's mass media. Most of them confidently announce that by the 1990s they – five to ten giant corporations – will control most of the world's important newspapers, magazines, books, broadcast stations, movies, recordings and videocassettes. Moreover, each of these planetary corporations plans to gather under its control every step in the information process, from the creation of 'the product' to all the various means by which modern technology delivers media messages to the public. 'The product' is news, information, ideas, entertainment and popular culture; the public is the whole world. (Bagdikian, 1989)

When Bagdikian wrote *The Media Monopoly* in 1983, he said that there were fifty corporations controlling the majority of media in the US. In 1987 he spoke of only twenty-five and the number is continuing to decrease not just in that country but throughout the world.

As never before, we are witnessing an astonishing concentration of capital. In 1979 the largest media merger in history took place when the Gannett chain of newspapers acquired a television company for a little more than US$350 million. Just nine years later, Rupert

Murdoch bought *TV Guide* and other magazines for the sum of US$3,000 million. And only eleven months after that the merger of Time Inc. and Warner Communications Inc. created the largest communications business in the world, valued at some US$18,000 million.

The First MacBride Round Table[6] (Harare, October 1989), which brought together communications experts and professionals from fourteen countries and eighteen non-governmental organizations, pointed out that:

> The debate on NWICO was not over one single issue but was related to the entire structure of world communication resources. It included such vital areas as international law, telecommunications, international trade and tariffs, transnational data flow, intellectual and artistic property rights, and the individual's right of privacy.   (Traber and Nordenstreng, 1992: 24–6)

In this area of new technologies, the private sector is exercising more and more control. All this is directly related to the second basic problem that faced the MacBride Commission: treating information as a commodity and not as a social asset.

The commoditization of information

We have reached a point at which knowledge itself has been turned into a commercial asset. Information is sold to whoever can pay for it. Two examples will make this clear. First, the right to literary ownership (copyright) is today the keystone of the global information economy. A legal concept aimed at regulating commerce in literary goods has been extended to the use of new technologies, especially in the area of computers. Today 'copyright' is not so much intended to benefit the public or the author, but the investor, which in the majority of cases means a transnational company.

Second, the current negotiations regarding GATT trade agreements and tariffs are not confined to 'free trade' but include control of knowledge. Privatizing information resources and scientific data are attempts at safeguarding intellectual property rights and patents. This will permit the transnational corporations to monopolize intellectual property. At the same time, all of this concerns the technological and industrial development of the Third World and will severely affect its medical and pharmaceutical services as well as its agricultural technology.[7]

Alongside the notion of social good with regard to communication, there are two other basic concepts to take into account: on the one hand, the educational process carried out through the mass media and, on the other, the receiver's right to participate in the communication process.

With respect to the first, the indifference of many countries to the educational capacity of the media is remarkable. Equally we can say that, in general, formal education remains on the fringe of the world of communication and regards television, for example, as an enemy of the school. As for the rights of the receiver, we should point out that it is not simply a matter of the misleading idea that the receiver's freedom lies in having a choice of newspaper to read or television programme to watch. What we are in fact experiencing is the opportunity to choose from an array of possibilities, not the freedom to choose what we really want.

A grave imbalance in the flow of information, TV programmes, films, magazines and books

Rapid developments in business communications in all spheres and their use by powerful firms is well known. A handful of news agencies controls no less than 80 per cent of world news. The greater part of news about what happens in Third World countries is received by them via the international centres and not from agencies belonging to the countries in which the news is generated. Thus, news about an event in Ghana goes first to London before being sent to neighbouring Nigeria. News about the Vietnam War was received in nearby Malaysia passing first through London and New York. Many local TV stations, because of a lack of technical and economic resources, simply reproduce what they get by satellite from the large agencies.

In the field of advertising, 'thirty-six of the 50 largest advertising companies in the world have their headquarters in the US. That is the equivalent of 19 of the 50 largest industrial companies.'[8] In Spain, Austria and Italy, seven of the twelve largest advertising companies belong to American companies.

Something similar occurred in films. To take just one example, in 1992 in Spain 2,008 films were shown of which only 384 were Spanish or co-productions (19.12 per cent), while 990 came from the US (49.30 per cent) and the rest from various European countries.[9]

Names like Murdoch, Maxwell, Bertelsmann and Berlusconi are heard almost daily in the media. Of course, they are the visible faces of a new power, the empire of information and communication with its concentration of television chains, newspapers of all kinds, publishing houses, and so on.

National sovereignty and international communications

Concentration of power in the hands of conglomerates on the strength of the information they possess thanks to the complex modern world of technology allows access to an enormous data bank, a large part of which is managed by multinational companies for their own advantage. This notion is related to the fifth problem that the MacBride Commission had to consider, which is perhaps the most prickly one: the sovereignty of different countries in the face of foreign interference through information and communications.

A technician from a multinational company with its headquarters in a Latin American country expressed surprise at the limited access he could gain to information in the data bank at the parent company. He knew that the information to which he could gain access was not the whole thing. He knew that all the data that he entered in the local computer was being recorded in the computer of the head office. What he did not realize was that he was only receiving the information that head office wanted him to have, yet he had to provide all the information that he had generated. A modern brain-drain had been effected at low cost by means of an electronic device.

Today we can understand the dimensions and complexities of this problem by making reference to two recent events.

In the first place, the Gulf War experience has produced a qualitative change at the global level. For the first time in the history of humanity, we have experienced what control of communications at the global level really means, and we saw that it has unpredictable consequences. The military-political conglomerate established rules of communication and the media had to fall in with them. For so-called reasons of security, the principles of free flow of information were temporarily suspended. Freedom of expression was controlled for security reasons in terms of protection and preservation. It seems reasonable to assume that protection was considered more important than freedom of information. But, what were they trying to protect?

We experienced communication which was used to create a fantasy instead of to inform the public of the true situation. References to the horrors of war were avoided. The aerial bombardments were 'surgical' and the bombs only fell on 'military targets'. Why this language? Maybe because triumphalism is insensitive to the pain of the defeated and also because 'politics is the art of preventing people participating in matters that concern them' (Paul Valéry). According to Philip Knightly the enforced censorship added a new dimension, 'it changed the public perception of the very nature of war' (Knightly, 1991: 4–5). William Fore reminds us that it was not by chance that from the day the US began to prepare for war up to when it started, television produced 2,855 minutes' coverage of the situation in the Gulf, of which just 29 (about 1 per cent) were devoted to opposition to this military venture (Fore, n.d.).

All this, of course, comprises one of the key issues in international communication of the 1990s: the development of a global surveillance system. Its origins go back to the beginnings of the post-1945 period and today, thanks to enormous developments in technology, a new theory of security has been set in motion.

Sandra Braman (1993: 36–40) says that the new theory of security is based on five factors:

- The geopolitical frontiers of nations have lost their importance for purposes of national security.
- The notion of national security has been extended beyond the military sphere to include the commercial and penal.
- The distinction between public and private spheres has been removed.
- The new theory, highlighting the ephemeral nature of defence, emphasizes recalling and processing information and the development of organizational forms to carry it out.
- The new theory of security rests in particular on the global infrastructure of information, especially the global surveillance system.

Communication in the service of community

This historical review would not be complete without reference to another aspect of the process – the development of what is known as popular or alternative communication.

Alternative communication emerged in the 1950s and developed in the 1970s and 1980s as a way of questioning traditional ways of doing communication and with the aim of placing communication at the service of the people. A good example of this is the development, from 1952 onwards, of miners' radio stations in Bolivia, at a time of profound social and economic change. The stations are financed by contributions from the meagre salaries of miners belonging to unions. In a little less than twelve years, they managed to set up twenty-seven broadcast stations which are noted for being self-managing, pluralist and, above all, of a participatory nature, giving people the opportunity to do their own communicating. In addition, there is another Bolivian example, 'dawn peasants' radio', an initiative of Aymara farmers who use time on commercial radios very early in the morning to broadcast in their own language and play their own music.[10]

What was being questioned? Communication, especially mass communication, for being vertical, and unidirectional in its sending of messages. The fact that this kind of communication is in the hands of a few whose basic aim is their own profit. That the commercial media, so powerful and far-reaching, are basically at the service of a consumerism in which human beings are considered *en masse*, not as beings with potential for freedom and creativity but as consumers who must be persuaded to consume and who have to understand that consumerism satisfies their essential needs. Alternative communication theory questioned the fact that traditional communication had to be 'massive', not in the sense of being received by large numbers of people but because of its influence on aspects that sustain the absence of freedom to act: passivity, inertia, weakening of the ability to think and decide. It questioned the fact that mass communication generally demonstrated lack of respect for local cultures, ignorance of and disinterest in the realities that the powerful countries were helping to maintain, and that its objective was cultural, economic and political domination.

It is important to point out that this questioning was part and parcel of the thinking that emerged at the beginning of the 1960s and became known throughout Latin America. In Venezuela, Antonio Pasquali was one of its instigators (Pasquali, 1963, 1980 and 1991), together with the Brazilian teacher Paulo Freire (Freire, 1973, 1976; Taylor, 1993). It would take too much space to detail the enormous contribution made by specialists in various countries, especially during the 1970s, to understanding media reality and the hegemonic influence of the industrialized countries on the non-industrialized (then called under-developed) and which led to the formation of national communication policies.

All of this work was supported by regional organizations that, even in the 1950s, were carrying out communication programmes:

organizations such as ILET (Instituto Latinoamericano de Estudios Transnacionales) and CREFAL (Centro Regional de Educación Fundamental de América Latina), both with headquarters in Mexico. At the end of that decade, CIESPAL (Centro Interamericano de Estudios Superiores de Periodismo para América Latina) was established in Ecuador, which would become the first centre on the continent to specialize in communications. Later, in the 1970s, another different ILET started up in Mexico, which did important work for the NWICO movement. Lastly, IPAL (Instituto para América Latina) began in Peru, initially dedicated to working on television and then on video. Finally, we should not forget the contribution made by Christian groups working in this field: the three Catholic organizations UNDA (Asociación Católica para la Radio, Televisión y Medios Afines), OCIC-ALM (Organización Católica Internacional de Cine para América Latina) and UCLAP (Unión Católica Latinoamericana de Prensa), among others, working in group communications, and WACC-LA/C working for liberation communication.

Alternative communication tried to focus on participation and dialogue, promoting people's reflection about their own reality; teaching them to express themselves through media; and providing adequate access to information. It insisted on the social character of media ownership. It stressed the critical recovery of wisdom and popular culture; recognition that the public has a voice and that it is important to listen to it; making the people subjects and protagonists of their own communication.

These ideas generated a way of working that characterized many social movements and which soon established a close relationship with popular education (Bordenave, 1987; Torres, 1989). Participation and training of popular groups multiplied. Two examples are video in Brazil, with the creation of the 'Association for Video Use by Popular Organizations' and training for peasants in Chile and Peru. Today there are networks covering practically the whole continent, sharing their work and co-operating for greater and better use of a medium that has proved highly popular (Gutierrez, 1989).[11] The alternative press has also played an important role. Its clearest example comes from Brazil during the time of its prolonged dictatorship. Called the 'midget press' because of its tiny formats and print-runs, it gave people the chance to communicate that which the commercial media did not allow.

The premise on which these activities was based was not just questioning the traditional system but attempting to supplant it. Alternative communication, in many places, began by deprecating mass communication and the new technologies. It took refuge in artesan management of communications; it privileged action over reflection; it increased the very limits of its messages; it tended towards atomization, isolation and lack of co-ordination of its communication experiences.

Today many of the protagonists of this movement are rethinking their ideas in the light of past experience and of the new realities they are facing. Beginning with the 1980s there was a current of renewal which proposed that communication has to understand itself from the basis of practice rather than theory and that the area of that practice is much wider than the world of the mass media. The media are not omnipotent and their audiences are not passive. Thus Jesús Martín-Barbero, from Columbia, insists that analysis must move 'from the media to mediations' (Martín-Barbero, 1987). and Néstor García Canclini emphasizes the reality of 'hybrid cultures' and the need to study 'together what is cultured, what is popular, what is mass, and the willingness to think again about the work of anthropology, sociology, art history, folklore and communication studies, which used to be kept separate' (Canclini, 1992: 11).

The time has certainly come to learn 'the lessons of the past'. It is at this point that global concerns coincide with ideas of community. Both have occupied a special place in the history of humanity. In order to study the critical situation facing communications ten years after the appearance of the MacBride Report, a group of communicators meeting in Lima, Peru, in 1990, declared among other things:

> Today more than ever before, emphasis on action rather than words means that we must seek a new way of communicating, which does not make myths of formulas and slogans, which does not ignore changes. At the same time we should not compromise the supreme ideal of a communication whose fundamental aspect is freedom from economic and political interests, and which is also more participatory and subject to higher criteria of solidarity and justice . . .   (Goicochea, 1991: 103)[12]

It may be that, among others, the following four ideas are paths in the search for new ways of communicating.

(1) Revalorization of communication media as potential spaces for popular communication and the need to promote structures in society which allow this to happen. Revalorizing the kind of mass

communication that uses the image, the form, the story, which relates to the senses, has global social impact and is a constituent part of the culture of these times.

(2) The need to recognize the tension between global and local, the group and the individual. A world of global communication is witnessing an upsurge of local cultures. The dream of homogenization ends in a plurality of viewpoints. The search for local expressions, the rescuing of indigenous cultures, the need to define one's own identity and roots are more and more apparent. All this leads to revalorization of the need for openness and inter-relationships. The development of communication networks in communities will inevitably increase, but we must also recognize and value the person and his or her needs and aspirations. Communication must be at the service of integration, dialogue and mutual respect.

At the same time we must recognize that communication interacts with economic and political development. We must consider the role of governments and their limits at the same time as that of the great economic powers that shape the destiny of communication and culture without even debating it but simply by placing it under the umbrella of so-called market forces.

(3) Recognition of the decisive role played by new technologies in communication. On the one hand, awareness of the rapid concentration by enormous government, military and commercial conglomerates thanks to high technology, of the flow of information and communication at the global level, which serve to increase disparity and dependency. The need to confront this new reality with a call for international regulations that limit ownerships and functions and establish to whom they are responsible.

On the other hand, taking advantage of these new technologies (computers, videos, e-mail) to set up new channels of international communication at the grassroots, creating new transnational networks to share information democratically.

(4) The need for communication to be recognized as a human right. The right of the individual to be the subject and not the object of communication. The right of popular groups to participate in the production and distribution of their own messages. The right to training, the development of communication capacities, especially in marginalized groups. The right for local cultures to be protected – those produced by the people – and for local cultures not to be

subordinated to commercial interests or to the great powers. The right to the free expression of ideas, encouraging media use by those who usually have no access to media. The right for communication services that are set up to be placed at the service of the integral development of the community, with a sense of participation and growth in the life of the community, which will necessarily be translated into a more just distribution of media ownership.

The search for authentic communication seems more like a Utopian undertaking and there are many reasons for having no faith in such a possibility. However, we should not forget the words of Simón Bolivar: that there is nothing more practical than a Utopia.

Notes

1   Gifreu (1986) is important for an understanding of the historical process, the debates and main ideas.
2   MacBride (1980). 'Unesco must guarantee the free flow of information full stop, said Mayor . . . He said plans for the New World Information Order "no longer exist" at Unesco, adding that they had violated human rights clauses in Unesco's charter' (*Washington Post*, 25 February 1989). Quoted by Roach (1990), p. 287.
3   A British Parliamentary committee seriously considered the possibility of returning to the bosom of Unesco, convinced that the economic arguments concerning the support Britain would have to provide if it rejoined were not relevant and saying that the Government had only been following in the footsteps of the US. (London: *Financial Times*, 4 August 1993, p.6).
4   H. I. Schiller, 'Forgetful and short-sighted – what hope for the future?' in *Media Development*, 3/1990. This issue included a series of reflections on Unesco's Medium-Term Plan, many of which expressed serious doubts about the future of the NWICO in Unesco.
5   See the important article by Annebelle Sreberny-Mohammadi (1991) on 'The Global and the Local in International Communications'. Equally Arjun Appadurai (1990), 'Disjuncture and Difference in the Global Cultural Economy', in which he insists that we should not see the new economy of global culture in terms of centre-periphery models but from five dimensions that he calls landscapes of ethnicity, the media, technology, finance and ideology.
6   The MacBride Round Table has met on four occasions and issued statements. The first three (Harare, Prague, Istanbul) were published in Traber and Nordenstreng (1992). The fourth (São Paulo) can be found in *Media Development* 2/1993 (pp. 48–9).
7   It is interesting to note that the US Ambassador to Argentina, James Cheek, recognized that 'lobbying is my mission' and that his main

objective was the approval of the medicine patent law. Quoted in the newspaper *La Nación*, international edition, Buenos Aires, 26 July 1993 (p. 3).

8 Karl Sauvant, quoted by Janus (1987), p. 121.

9 *Cine Informe*, 32, 637, Madrid, May 1993 (p. 5).

10 Tealdo (1989), especially the chapters 'The Bolivian Miners' Radios' by Teresa Flores, and 'Peasant Intervention in Bolivian Radio Broadcasting' by Sandra Aliaga and Magali Camacho. For other Peruvian examples, see CEPES (1987). There is a good series of articles on this theme in *Media Development* 4/1990.

11 See also the issue on 'Video for the people', *Media Development* 4/1989.

12 Other important contributions are to be found in Fox (1988).

References

Appadurai, Arjun (1990). 'Disjuncture and difference in the global cultural economy', *Public Culture*, Vol. 2, 2. London.

Bagdikian, Ben (1983). *The Media Monopoly*. Boston MA: Beacon Press.

Bagdikian, Ben (1989). 'The lords of the global village', *The Nation* 43, Vol. 248, New York, June 1989.

Bordenave, Juan Díaz (1987). 'Comunicación y Educación', in *Comunicación y Desarrollo*. Lima: IPAL.

Braman, Sandra (1993). 'Global surveillance, media politics and civil liberty', in *Media Development* 2/1993.

Canclini, Néstor García (1992). *Culturas Híbridas*. Buenos Aires: Sudamericana.

CEPES (1987). *Radio y Comunicación Popular en el Perú*. Lima: CEPES.

Estévez, Jaime (1993). *Crisis del Orden Internacional y Tercer Mundo*. Mexico: CEESTEM-Nueva Imagen.

Fore, William (n.d.). 'The shadow war in the gulf'. Unpublished manuscript.

Fox, Elizabeth (ed.) (1988). *Media·Politics and Latin America*. London: Sage.

Freire, Paulo (1973). *Education for Cultural Consciousness*. New York: Seabury Press.

Freire, Paulo (1976). *Education, the Practice of Freedom*. London: Writers' and Readers' Co-operative.

Grifeu, Josep (1986). *El debate internacional de la comunicación*. Barcelona: Ariel.

Guía del Tercer Mundo (1988). Buenos Aires: Colihue.

Gutierrez, Mario (ed.) (1987). *Video, Tecnología y Comunicación Popular*. Lima: IPAL.

Goicochea, Pedro (ed.) (1991). *América Latina: Las comunicaciones cara al 2000*. Lima: IPAL.

Janus, Noreene (1987). 'Propaganda, medios de comunicación masivos y la formación de una cultura en el Tercer Mundo', in *Publicidad: la otra cultura*. Lima: IPAL.

Knightly, Philip (1991). 'Here is the partially censored news', in *Index on Censorship*, 4 and 5. London.

MacBride, Seán (1980). *Many Voices, One World*. Paris: Unesco.

McLuhan, Marshall and Quentin Fiore (1987). 'The new electronic interdependency recreates the world in the image of a global village', *The Medium is the Massage*. New York: Bantam.

Martín-Barbero, Jesús (1987). *De los medios a las mediaciones: Comunicación, cultura y hegemonía*. Mexico: Gili. Translated as *Communication, Culture and Hegemony*, London: Sage (1993).

Pasquali, Antonio (1963). *Cultura de Masas*. Caracas: Monte Avila.

Pasquali, Antonio (1980). *Comprender la Comunicación*. Caracas: Monte Avila.

Pasquali, Antonio (1991). *El Orden Reina*. Caracas: Monte Avila.

Roach, Colleen (1990). 'A New World Information and Communication Order', *Media Culture and Society*, July 1990.

Schiller, H. I. (1990). 'Forgetful and short-sighted – what hope for the future?', *Media Development*, 3/1990.

Sreberny-Mohammadi, Annebelle (1991). 'The global and the local in international communications', in James Curran and Michael Gurevitch (ed.), *Mass Media and Society*. London: Edward Arnold.

Taylor, Paul V. (1993). *The Texts of Paulo Freire*. Philadelphia: Open University Press.

Tealdo, Ana Rosa (1989). *Radio y Democracia*. Lima: IPAL.

Traber, Michael and Kaarle Nordenstreng (1992). *Few Voices, Many Worlds*. London: World Association for Christian Communication.

Torres, Rosa María (1989). 'Educación popular y comunicación popular', in *El video en la educación popular*. Lima: IPAL.

Unesco (1993). *Recommendations of the Executive Board on the Draft Programme and Budget for 1994–1995*, 27 1/6. Paris: Unesco.

Index

accountability, principle of 19–20, 23–4, 29
in journalism 121–2, 127
accuracy, in journalism 115–16
advertising agencies 207
agrarian societies 47–9
alternative communication 139, 209–14
anticipatory modernization 47–9
Ashley, Wayne 148
Augustine of Hippo, Saint 86–7
autonomy, individual 78–82
as principle of Enlightenment 78–82
in religious matters 193, 194
autonomy, journalistic 117–18

balance of power 61
media–society relations 119
printing and 189–90
Basavanna 145
Baskin, William 136–7
Basque language 177, 182
Bolivia, community radio services 209–10
book publishing, Europe 75–6
Braman, Sandra 209
Brazil
industrial development 200
popular communication 211
Bretton Woods Agreement (1944) 55, 59
broadcasters see journalists/broadcasters
Breton language 176–7
Broadcasting & Cable (trade paper) 163, 166, 167

Callahan, Daniel 82
capitalism
and participatory democracy 16, 24, 31–3
as future world order 204
development of 41, 46, 48–50, 52–4, 68–9
liberal 55–6, 69, 72
vulnerability of 60
Carey, John 131
Carter, Stephen 188–9
casting and fate, television 156–8
Catalan language 177, 181
CEM (Cultural Environment Movement) 170
censorship 29–31
Gulf War 208–9
CI (Cultural Indicators) project 155–63, 168
city states 49–51
civic/public journalism 120–1
Cold War
and journalism 123, 127
ending of 58–9; see also Communism, collapse of
collective communication 96–7
command economies see Communism
commoditization of information 206–7
commodity fetishism 62–3
'common human heritage' 21
communication
access strategies 67, 190–2
as medium for education 4–8, 207
cultural and discursive 186–8
definition 94
economic development and 199–214
national sovereignty and 208–9
public control of 92
relationship with democracy 15–37, 143–54
right of 95–7, 213–14
social movements and 92–113

traditional 143–54
communication channels/resources,
ownership of 22, 128, 169, 205–6
communication élites 39–40, 54, 56–7, 64
communication ethics 75–91, 105; *see
also under* journalists/broadcasters
communication technology
and democracy 130–1
use by social movements 106
women and 130–42
see also technological development
communication theory, nature and
limitations of 16–18
Communism, collapse of 15, 31, 42, 58, 204
and journalistic professionalism 127
and ethno-nationalism 52, 62
Communism, development of 53–4, 55–6
community, growth and maintenance of
150–3, 213
community-based communication
187–8, 196, 209–14
communologues 57, 64
consciousness raising 109–10
constitutional democracy 52–3
Consultative Club 124–7
consumerism 205, 210
consumer power 24
copyright/intellectual property 21–2, 206
Counter Enlightenment 84
credibility, religion 192
cultural dialogue 98–9
cultural environment
as a creation of the media 187
as a given, in need of protection
164, 169–70
Cultural Environment Movement *see*
CEM
cultural frontier of democracy 155–74
Cultural Indicators (CI) project 155–63,
168
cultural relativism 58
cyberspace 60
cybermedia 63–4
gender aspects 138

D'Arcy, Jean 6
decision-making, democratic 19, 22–3
Declaration of Persepolis (1975) 5
Declaration of the Rights of Man and of
the Citizen (1789) 5
Delhi Declaration (1993) 4
democracy and the Enlightenment 75–91
democratic communication, concept of
93–4

democratic compromise 1–2
democratic ideal 15–37
democratization, global sequence of
38–74
Descartes, René 77–8
de Toqueville, Alexis 81
dichotomies of the Enlightenment 75–82
discursive environment 187–8, 196

Eastern Europe *see* Communism,
collapse of
economic democratization 65–72
economic development, and
communication 199–214
economics, and liberty 30
education
and communication 4–8, 207
North/South divide 4, 6
relationship with ideology 3–4
see also media education
egalitarianism, justification of 25–31
Eisenstein, Elizabeth 189
empowerment 18, 20–2, 109–10
barriers to 31
Enlightenment 75–82
and modern-day science 135
mathematics 76–8, 80
paradigm for the press 83–4
romantic vs. rational 79–80
epistemology of news 85
ethical principles/values 8–10, 75–91, 105
democracy and 81–2
journalism 13–14, 84, 115–16,
121–7
normative theory 17
Protestant work ethic 52
universality of 26–7, 82, 88, 125,
147–8
ethno-nationalism 62
ethno-religious movements 57, 59, 62, 69
European Commission on Human Rights
25
European institutions, and minority
languages 178
exclusivism 100, 145, 150

federation and subsidiarity 107–8
feminist issues 130–42
aspects of communication
technology 137–9
'feminist rationalization for failure'
136
feminist standpoint theory 135
liberal vs. radical 137

finance capitalism 53
fortress journalism 118, 120, 126–7
Fourth and Fifth Worlds 43
Fox Broadcasting 165
France, minority languages 176–8
free flow of information 202–3
    suspension of 208–9
freedom, principle of 12, 18–19, 78–82
    and the moral order 79, 80–2
    and responsibility 83
    economics and 30
    need for religion 81
    of conscience 49
    of information 95, 120, 202–3,
        208–9
    of the press 203
    of speech 29–31, 72, 100
Freedom Forum Media Studies Center
    (New York) 130–1
Freire, Paulo 106–7, 109, 210
Frisian language 177

Gaelic, Scottish and Irish 177–8
Galileo Galilei 76, 78
Gandhi, M. K. 147
GATT/WTO 21–2, 55, 58–9
    and copyright 206
    as destabilizing influence 59
gender-related issues 130–42
    in communication technology
        137–9
    in science 135–7
    in television programming 156–60
General Agreement on Tariffs and Trade
    see GATT/WTO
genre creation 98, 105–6, 108
GII (Global Information Infrastructure)
    15–16
global censorship 208–9
global citizenship 42
Global Information Infrastructure *see* GII
global perspectives on democracy 38–74
'global village' (McLuhan) 199–200
globalization of markets 31–3, 42, 44,
    55–6, 58–60, 62, 207–8
    for US television output 158–9,
        166–70
'glocalization' 60–1
Gore, A. 15
'Group of 77' 201
Gulf War 208–9

Habermas, J. 26–7, 146–7
Hamelink, Cees 119

Harding, Sandra 132–3
Harvey, Ethel Browne 134
historical conceptualizations of
    democracy 38–9
historical watersheds 40–65
Holloway, Margaret 133–4
Hollywood Caucus of Producers, Writers
    and Directors 165
Hooke, Robert 80
horizontal communications networks
    102–3
human dignity 10–13
human rights 5, 6, 12, 48
    and remedies 24–5
    as universal principle 28–9
humanism 49
Hutchins Commission 116
hyperspace 65

IBRD (International Bank for
    Reconstruction &
    Development/World Bank) 55, 59
identity fetishism 63
identity fusion and differentiation 42–3
ideologies of discrimination 71
ideologues 54
IFJ (International Federation of
    Journalists) 124, 127–8
ILO (International Labour Organization)
    23, 122
IMF (International Monetary Fund) 55, 59
imperial systems 47–9
India
    culture 146
    folk theatre 148–9, 151
    women's movement 139
individual autonomy *see* autonomy,
    individual
individual goals 95
individual responsibility 49
industrial imperialism 53–4; *see also*
    multinational corporations
information
    commodization of 206–7
    right to 95, 120
    right to contribute to 96, 213
information superhighway 2, 15, 130–1,
    139–40
'infotainment' 117
instrumental rationality 97–8
instrumental relations 186
Integrated System Digital Network *see*
    ISDN
intellectual property/copyright 21–2, 206

interdependence, in social movements
102–3
International Bank for Reconstruction &
Development/World Bank *see* IBRD
International Federation of Journalists
see IFJ
International Labour Organization *see*
ILO
International Monetary Fund *see* IMF
International Organization of Journalists
see IOJ
International Principles of Professional
Ethics in Journalism 13–14
intersubjectivity, principle of 27–8
IOJ (International Organization of
Journalists) 123, 124, 127–8
ISDN (Integrated System Digital
Network) 56

jestologues 64
Johnson, Mark 85
journalists/broadcasters 114–29
accountability 121–2
autonomy 177–18
civic/public journalism 120–1
ethical principles 13–14, 84, 115–16,
121–7
historical perspective 122–7
journalistic paradox 115–18
paradigm 117, 121
self-regulation 121–2, 127
training of 7, 117
justice, principle of 12–13
justification of egalitarianism 25–31

language
and truth 88
nature and role of 88–9
see also linguistic minorities;
linguistic nationalism
Latin America
Christian organizations 211
community-based communication
209–11
popular culture 149–50, 211, 212
regional educational and religious
organizations 211
legislation and regulation, media 170,
175, 213
self-regulation, journalism 121–2,
127
liberal nationalism 52
libertarianism 30–1
liberty *see* freedom

libraries, ancient 47
linguistic minorities, and the media
175–84
effect of news agencies 182–3
links with group
consciousness/culture 179–80
linguistic nationalism 176–9
literacy 5, 76
in minority languages 180
local communities
community-based communication
187–8, 196, 209–14
involvement in global affairs 34

MacBride Commission (1977–80) 10,
114, 122, 124, 199, 202
problems addressed by 205–9
McLuhan, Marshall 199–200
manufacturing capitalism 53
Many Voices, One World (MacBride
Report, 1980) 114, 122, 124, 202
market economies *see* capitalism
mass media
access to 67, 96, 175, 190–2
and democracy 175
and identity fetishism 63
concentration in ownership of 22,
128, 169, 205–6
creation of 'global village' 199–200
democratic administration of 107–8
development of 53, 54
genre creation 98, 105–6, 108
linguistic minorities and 175–84
Mass Media Declaration, 1978
(Unesco) 124, 126
media–society relations 118–22
propaganda use of 56, 108–9, 175
regional 183
regulation of 121–2, 127, 170, 175,
213
religious pluralism and 185–98
see also journalists/broadcasters;
press; public service media; radio;
television
mathematics 76–8, 80
media education, role of 6–7, 9
media legislation and regulation 170,
175, 213
self-regulation, journalism 121–2,
127
media reform movements 92
media-society relations 118–22
mercantile capitalism 50–1
Milton, John 100

modernism, origins of 75, 83
modernization
    and religion 185–98
    global sequence of 38–74
    processes of 145–6
morality *see* ethical principles
MTV 64–5
multinational corporations 205–8

nation states 51–3, 60
    linguistic and political nationalism
    176–9
    traditions of 145
national sovereignty, and international
    communications 208–9
New International Economic Order *see*
    NIEO
New World Information and
    Communication Order *see* NWICO
'new world order' 31–3, 59
newly industrialized countries *see* NICs
news agencies
    and linguistic minorities 182–3
    control of world news 207
Newton, Isaac 76–7
NICs (newly industrialized countries)
    43–4
Nielsen ratings, television 161–3
NIEO (New International Economic
    Order) 201–2
non-serious journalism 117
North/South divide
    and use of science and technology
    132–3
    closing, in journalism 121
    in access to mass media 199–200
    in education 4, 6
    industrial and economic
    development 200–2
NWICO (New World Information and
    Communication Order) 121, 201–3

objectivism 83–4
organic relations 186
Oriental philosophy/culture 143, 146; *see
    also* India
ownership of the media 22, 128, 169,
    205–6

para-church organizations 193–5
paradigms 100, 111–12
    of communication 204
    of journalism 117, 121
    see also public cultural truth

Parsons, Talcott 83–4
participation, principle of 13, 18–19,
    22–3, 29, 93, 95, 213–14
    alternative communication and 211
    participatory culture 104–7, 111–12
Pasquali, Antonio 210
peace, principle of 13
Peace of Westphalia (1648) 51
People's Communication Charter 5, 119
philosophes 75
plausibility, religion 191
political equality 18–20
political religion *see* ethno-religious
    movements
polls, on violence in entertainment 164,
    166
popular culture 106–7, 117, 143
    Latin America 149–50, 211, 212
postmodernism 27–8, 57–8, 194
poverty, as 'moral failure' 63
power
    and access to information 2–3,
    101–2
    balance of 61, 119, 189–90
    delegation of 19
    in ownership of media 22, 128, 169,
    205–8
    of the state, limitation of 24
    of technology 137–8
pragmatism, of traditional
    communication 151–2
press
    as Enlightenment paradigm 83–4
    freedom of 203
    see also journalists/broadcasters;
    mass media
printing, development of 50, 51, 186,
    189–90
private discursive communication 187, 196
professionalism
    dangers to democracy 114, 118
    in journalism 125–7
Protestant Reformation 49–51
Protestant work ethic 52
public accountability *see* accountability
public/civic journalism 120–1
public communication *see*
    communication
public cultural rituals 111
public cultural truth 93–5; *see also*
    paradigms
public discursive communication 188, 196
public philosophy of communication
    99–100

public service media
    ethical basis for 9
    privatization of 168
publicity, for religion 190–2

radio
    Bolivian community services
    minority languages on 177, 180, 210
rapidity, in journalism 116
Rawls, J. 26–7
Reformation 49–51
regionalism 60–1
    regional media 183
regulation of media *see* legislation and
    regulation
relational dynamics 148–9
religion
    as root of democracy 39, 47
    contemporary evolution 192–5
    ethno-religious movements 57, 59,
    62, 69
    Latin American Christian
    organizations 211
    national churches 51–2
    publicity for 190–2
    religious pluralism, and the media
    185–98
    religious-symbolic 'marketplace'
    194–5
    revivalism 60, 69
    televangelists 191, 194
remedies, access to 19, 24–5, 29
Renaissance 49–51
root paradigm 100, 111–12
Rorty, Richard 83, 85
Rosen, Jay 120
Rousseau, Jean-Jacques 78–9
Rushdie, Salman 64

Saturday morning television 157–60
'science question' 131–7
self-regulation, journalism 121–2, 127
seriousness, in journalism 116–17
sex, on television 169
Sikes, Alfred C. 130–1
small media, importance to social
    movements 109–10
social movements, and communication
    92–113, 209–14
    characteristics of social movements
    101
societal goals 38
space, collapse of 42, 46
space technology 21–2

Spain, minority languages 177, 181, 182
speech, freedom of 29–31, 72, 100
state capitalism 68
strategic periodicals 11
symbols/symbolic language, as
    expression of public cultural truth
    94, 103–4

technological development 20–2
    acceleration of 42, 46
    and communication 56–7, 131–7,
    199–214
    distribution of infrastructures 20
technologues 56, 64
technology transfer 20–2, 201
televangelists 191, 194
television, minority languages on 176–84
television programming, US 155–70
    casting and fate 156–8
    gender, class and ethnicity 156–60
    global marketing 158–9, 166–70
    Nielsen ratings 161–3
    sex 169
    themes 169–70
    violence 158–66
Third World *see* North/South divide
TNCs (transnational corporations) 55
totalitarianism 54, 63
Traber, Michael 10–14, 75, 89, 139–40, 155
trade unions, journalism 122–3, 128
traditional communication 143–54
    and community 150–3
    and religion 191
    and the status quo 148, 152
    dynamism of 149–50
    imagination and 152
    nature and forms of 144–5
    role of 149
transnational corporations *see* TNCs
truth
    and the moral order 85–8
    as norm 82–8, 89
    as reason radiated by *caritas* 86–7
    in journalism 115–16
    language and 88
    principle of truth-telling 12, 85
'tsunamis' in the democratization process
    40–65

Unesco 123–4, 127–8, 202–4
    Mass Media Declaration (1978)
    124, 126
    Third Medium-Term Plan (1990–5)
    203

see also MacBride Commission
United Nations 55
   World Conference on Human Rights
   (1993) 29
Universal Declaration of Human Rights
   (1948) 5, 6
universalization, principle of 26
universals 26–7, 82, 88, 125, 147–8
universities, development of 51, 57
US television 155–70
utilitarian philosophies 26
Utopian vision 214

vernacular publishing 75–6
violence, on television 158–66
Voltaire 77

WACC (World Association for Christian
   Communication) 10–12
Welsh language 175–7, 181
women's issues 130–42
   ethos of women's studies 136
   women in science 133–5
   women on US television 156–60
World Association for Christian
   Communication *see* WACC
World Bank/International Bank for
   Reconstruction & Development *see*
   IBRD
World Meetings of Journalists 123
World Trade Organization *see*
   GATT/WTO
writing, development of 47, 48